Programmable Logic Controllers

Programmable Logic Controllers

An Introduction

Second Edition

W. Bolton

 Newnes

OXFORD AUCKLAND BOSTON JOHANNESBURGH MELBOURNE NEW DELHII

Newnes
An imprint of Butterworth-Heinemann
Linacre House, Jordan Hill, Oxford OX2 8DP
225 Wildwood Avenue, Woburn, MA 01801-2041
A division of Reed Educational and Professional Publishing Ltd

A member of the Reed Elsevier plc group

First published 1996
Second edition 2000

© W. Bolton 1996, 2000

British Library Cataloguing in Publication Data
A catalogue record for this book is available from the British Library

ISBN 07506 47469

Library of Congress Cataloguing in Publication Data
A catalogue record for this book is available from the Library of Congress

Printed and bound in Great Britain

FOR EVERY TITLE THAT WE PUBLISH, BUTTERWORTH-HEINEMANN
WILL PAY FOR BTCV TO PLANT AND CARE FOR A TREE.

Contents

Preface

Technological advances in recent years have resulted in the development of the programmable logic controller and a consequential revolution of control engineering.

This book is an introduction to programmable logic controllers and aims to ease the tasks of practising engineers coming first into contact with programmable logic controllers and also provide a basic course for students on programmes such as Higher Nationals, company training programs and as an introduction for first year undergraduate courses in engineering.

Changes from the first edition

The main changes from the first edition are:

- More on the internal architecture of PLCs.
- More on input/output devices.
- More on input/output processing.
- More on networks.
- More on sequencers and continuous control.
- More on number systems, including problems.

and, as before:

- Lots of illustrations of how to program PLCs, whatever the manufacturer, and exercises to test whether the reader understands the principles and is able to write such programs.

The new edition has been designed to provide full syllabus coverage of the new Higher National unit in PLCs from Edexcel and the Advanced GNVQ Optional Unit in PLCs which forms part of the new specifications starting in the year 2000.

The book addresses the problem of different programmable control manufacturers using different nomenclature and program forms by describing the principles involved and illustrating them with examples from a range of manufacturers

Aims

This book aims to enable the reader to:

- Identify and explain the main design characteristics and internal architecture of programmable logic controllers.
- Describe and identify the characteristics of commonly used input and output devices.
- Explain the processing of inputs and outputs by programmable logic controllers.
- Describe communication links involved with PLC systems.
- Develop ladder programs for the logic functions AND, OR, NOR, NAND, NOT and XOR.
- Develop ladder programs involving internal relays, timers, counters, shift registers, sequencers and data handling.
- Identify methods used for testing and debugging.

Structure of the book

The following figure outlines the structure of the book and its relationship to the three unit outcomes of the BTEC specification:

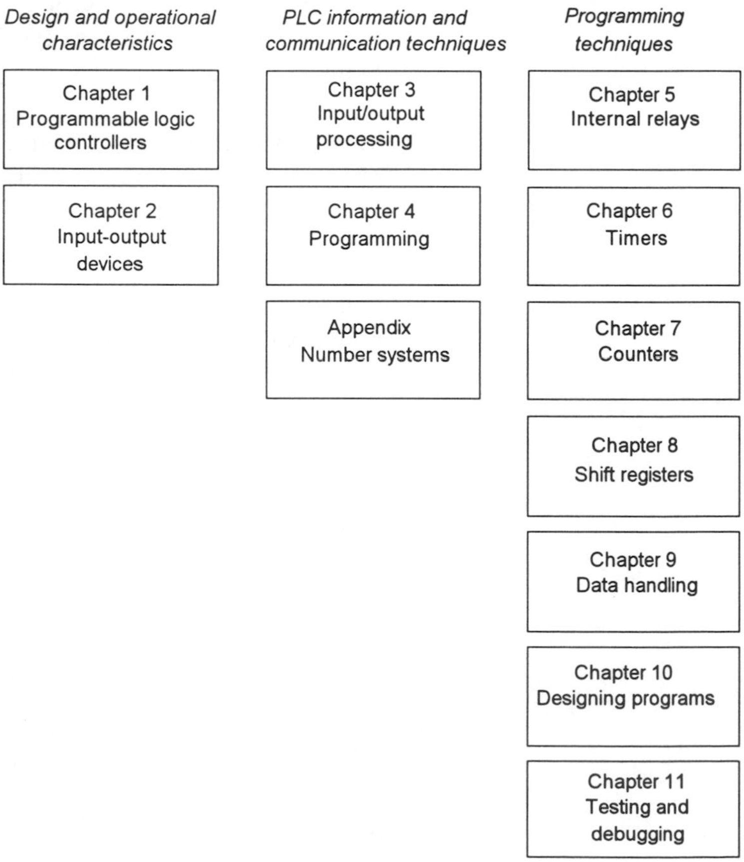

Design and operational characteristics	*PLC information and communication techniques*	*Programming techniques*
Chapter 1 Programmable logic controllers	Chapter 3 Input/output processing	Chapter 5 Internal relays
Chapter 2 Input-output devices	Chapter 4 Programming	Chapter 6 Timers
	Appendix Number systems	Chapter 7 Counters
		Chapter 8 Shift registers
		Chapter 9 Data handling
		Chapter 10 Designing programs
		Chapter 11 Testing and debugging

To assist the reader to develop the skills necessary to write programs for programmable logic controllers, many worked examples, multi-choice questions and problems are included in the book with answers to all multi-choice questions and problems given at the end of the book.

W. Bolton

1 Programmable logic controllers

This chapter is an introduction to the programmable logic controller, its general function, hardware forms and internal architecture. This overview is followed up by more detailed discussion in the following chapters.

1.1 Controllers

What type of task might a control system have? It might be required to control a sequence of events or maintain some variable constant or follow some prescribed change. For example, the control system for an automatic drilling machine (Figure 1.1(a)) might be required to start lowering the drill when the workpiece is in position, start drilling when the drill reaches the workpiece, stop drilling when the drill has produced the required depth of hole, retract the drill and then switch off and wait for the next workpiece to be put in position before repeating the operation. Another control system (Figure 1.1(b)) might be used to control the number of items moving along a conveyor belt and direct them into a packing case. The inputs to such control systems might be from switches being closed or opened, e.g. the presence of the workpiece might be indicated by it moving against a switch and closing it, or other sensors such as those used for temperature or flow rates. The controller might be required to run a motor to move an object to some position, or to turn a valve, or perhaps a heater, on or off.

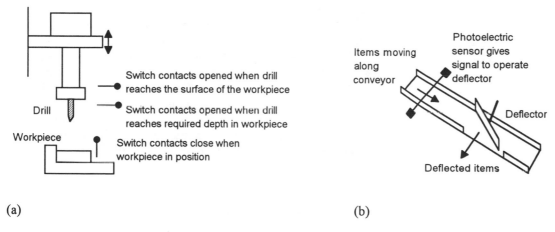

(a)   (b)

Figure 1.1 *An example of a control task and some input sensors, (a) an automatic drilling machine, (b) a packing system*

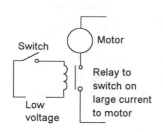

Figure 1.2 *A control circuit*

What form might a controller have? For the automatic drilling machine, we could wire up electrical circuits in which the closing or opening of switches would result in motors being switched on or valves being actuated. Thus we might have the closing of a switch activating a relay which, in turn, switches on the current to a motor and causes the drill to rotate (Figure 1.2). Another switch might be used to activate a relay and switch on the current to a pneumatic or hydraulic valve which results in pressure being switched to drive a piston in a cylinder and so results in the workpiece being pushed into the required position. Such electrical circuits would have to be specific to the automatic drilling machine. For controlling the number of items packed into a packing case we could likewise wire up electrical circuits involving sensors and motors. However, the controller circuits we devised for these two situations would be different. In the 'traditional' form of control system, the rules governing the control system and when actions are initiated are determined by the wiring. When the rules used for the control actions are changed, the wiring has to be changed.

1.1.1 Microprocessor controlled system

Instead of hardwiring each control circuit for each control situation we can use the same basic system for all situations if we use a microprocessor-based system and write a program to instruct the microprocessor how to react to each input signal from, say, switches and give the required outputs to, say, motors and valves. Thus we might have a program of the form:

> If switch A closes
> Output to motor circuit
> If switch B closes
> Output to valve circuit

By changing the instructions in the program we can use the same microprocessor system to control a wide variety of situations.

As an illustration, the modern domestic washing machine uses a microprocessor system. Inputs to it arise from the dials used to select the required wash cycle, a switch to determine that the machine door is closed, a temperature sensor to determine the temperature of the water and a switch to detect the level of the water. On the basis of these inputs the microprocessor is programmed to give outputs which switch on the drum motor and control its speed, open or close cold and hot water valves, switch on the drain pump, control the water heater and control the door lock so that the machine cannot be opened until the washing cycle is completed.

1.1.2 The programmable logic controller

A *programmable logic controller* (PLC) is a special form of micro-processor-based controller that uses a programmable memory to store

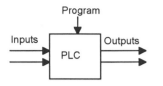

Figure 1.3 *A programmable logic controller*

instructions and to implement functions such as logic, sequencing, timing, counting and arithmetic in order to control machines and processes (Figure 1.3) and are designed to be operated by engineers with perhaps a limited knowledge of computers and computing languages. They are not designed so that only computer programmers can set up or change the programs. Thus, the designers of the PLC have pre-programmed it so that the control program can be entered using a simple, rather intuitive, form of language, see Chapter 4. The term *logic* is used because programming is primarily concerned with implementing logic and switching operations, e.g. if A or B occurs switch on C, if A and B occurs switch on D. Input devices, e.g. sensors such as switches, and output devices in the system being controlled, e.g. motors, valves, etc., are connected to the PLC. The operator then enters a sequence of instructions, i.e. a program, into the memory of the PLC. The controller then monitors the inputs and outputs according to this program and carries out the control rules for which it has been programmed.

PLCs have the great advantage that the same basic controller can be used with a wide range of control systems. To modify a control system and the rules that are to be used, all that is necessary is for an operator to key in a different set of instructions. There is no need to rewire. The result is a flexible, cost effective, system which can be used with control systems which vary quite widely in their nature and complexity.

PLCs are similar to computers but whereas computers are optimised for calculation and display tasks, PLCs are optimised for control tasks and the industrial environment. Thus PLCs are:

1 Rugged and designed to withstand vibrations, temperature, humidity and noise.
2 Have interfacing for inputs and outputs already inside the controller.
3 Are easily programmed and have an easily understood programming language which is primarily concerned with logic and switching operations.

The first PLC was developed in 1969. They are now widely used and extend from small self-contained units for use with perhaps 20 digital inputs/outputs to modular systems which can be used for large numbers of inputs/outputs, handle digital or analogue inputs/outputs, and also carry out proportional-integral-derivative control modes.

1.2 Hardware

Typically a PLC system has five basic components. These are the processor unit, memory, the power supply unit, input/output interface section and the programming device. Figure 1.4 shows the basic arrangement.

1. The *processor unit* or *central processing unit (CPU)* is the unit containing the microprocessor and this interprets the input signals and carries out the control actions, according to the program stored in its memory, communicating the decisions as action signals to the outputs.

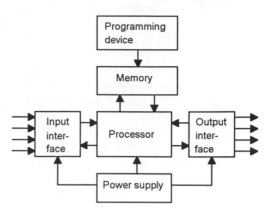

Figure 1.4 *The PLC system*

2   The *power supply unit* is needed to convert the mains a.c. voltage to the low d.c. voltage (5 V) necessary for the processor and the circuits in the input and output interface modules.

3   The *programming device* is used to enter the required program into the memory of the processor. The program is developed in the device and then transferred to the memory unit of the PLC.

4   The *memory unit* is where the program is stored that is to be used for the control actions to be exercised by the microprocessor.

5   The *input and output sections* are where the processor receives information from external devices and communicates information to external devices. The inputs might thus be from switches, as illustrated in Figure 1.1(a) with the automatic drill, or other sensors such as photo-electric cells, as in the counter mechanism in Figure 1.1(b), temperature sensors, or flow sensors, etc. The outputs might be to motor starter coils, solenoid valves, etc. Input and output interfaces are discussed in Chapter 2. Input and output devices can be classified as giving signals which are discrete, digital or analogue (Figure 1.5). Devices giving *discrete* or *digital signals* are ones where the signals are either off or on. Thus a switch is a device giving a discrete signal, either no voltage or a voltage. *Digital* devices can be considered to be essentially discrete devices which give a sequence of on–off signals. *Analogue* devices give signals whose size is proportional to the size of the variable being monitored. For example, a temperature sensor may give a voltage proportional to the temperature.

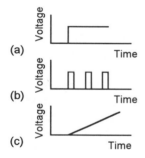

Figure 1.5 *Signals: (a) discrete, (b) digital, (c) analogue*

1.2.1 Mechanical design of PLC systems

There are two common types of mechanical design for PLC systems; a *single box*, and the *modular* and *rack types*. The single box type is commonly used for small programmable controllers and is supplied as an integral compact package complete with power supply, processor, memory,

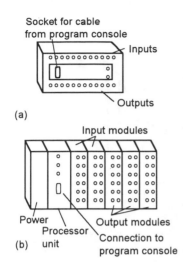

Socket for cable
from program console

Inputs

Outputs

(a)

Input modules

Power

Processor

(b) unit

Output modules

Connection to
program console

Figure 1.6 *(a) single box, (b) modular/rail type*

Screen

Labelled keys
for entering
the program

Figure 1.7 *Hand-held programmer*

and input/output units (Figure 1.6(a)). Typically such a PLC might have 40 input/output points and a memory which can store some 300 to 1000 instructions. The modular type consists of separate modules for power supply, processor, etc. which are often mounted on rails within a metal cabinet. The rack type can be used for all sizes of programmable controllers and has the various functional units packaged in individual modules which can be plugged into sockets in a base rack (Figure 1.6(b)). The mix of modules required for a particular purpose is decided by the user and the appropriate ones then plugged into the rack. Thus it is comparatively easy to expand the number of input/output connections by just adding more input/output modules or to expand the memory by adding more memory units.

Programs are entered into a PLC's memory using a program device which is usually not permanently connected to a particular PLC and can be moved from one controller to the next without disturbing operations. For the operation of the PLC it is not necessary for the programming device to be connected to the PLC since it transfers the program to the PLC memory.

Programming devices can be a hand-held device, a desktop console or a computer. Hand-held systems incorporate a small keyboard and liquid crystal display, Figure 1.7 showing a typical form. Desktop devices are likely to have a visual display unit with a full keyboard and screen display. Personal computers are widely configured as program development workstations. Some PLCs only require the computer to have appropriate software, others special communication cards to interface with the PLC. A major advantage of using a computer is that the program can be stored on the hard disk or a floppy disk and copies easily made. The disadvantage is that the programming often tends to be not so user-friendly. Hand-held programming consoles will normally contain enough memory to allow the unit to retain programs while being carried from one place to another.

Only when the program has been designed on the programming device and is ready is it transferred to the memory unit of the PLC.

1.3 Internal architecture

Figure 1.8 shows the basic internal architecture of a PLC. It consists of a central processing unit (CPU) containing the system microprocessor, memory, and input/output circuitry. The CPU controls and processes all the operations within the PLC. It is supplied with a clock with a frequency of typically between 1 and 8 MHz. This frequency determines the operating speed of the PLC and provides the timing and synchronisation for all elements in the system. The information within the PLC is carried by means of digital signals. The internal paths along which digital signals flow are called *buses*. In the physical sense, a bus is just a number of conductors along which electrical signals can flow. It might be tracks on a printed circuit board or wires in a ribbon cable. The CPU uses the *data bus* for sending data between the constituent elements, the *address bus* to send the addresses of locations for accessing stored data and the *control bus* for signals relating to internal control actions. The *system bus* is used for communications between the input/output ports and the input/output unit.

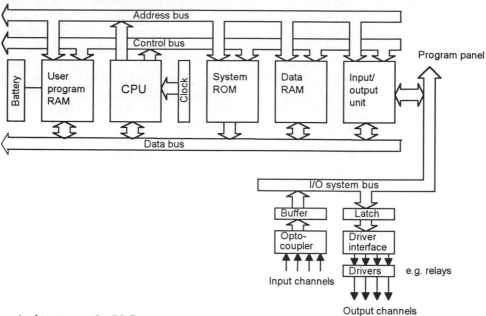

Figure 1.8 *Architecture of a PLC*

1.3.1 The CPU

The internal structure of the CPU depends on the microprocessor concerned. In general they have:

1 An *arithmetic and logic unit* (ALU) which is responsible for data manipulation and carrying out arithmetic operations of addition and subtraction and logic operations of AND, OR, NOT and EXCLUSIVE-OR.
2 Memory, termed *registers*, located within the microprocessor and used to store information involved in program execution.
3 A *control unit* which is used to control the timing of operations.

1.3.2 The buses

The buses are the paths used for communication within the PLC. The information is transmitted in binary form, i.e. as a a group of *bits* with a bit being a binary digit of 1 or 0, i.e. on/off states. The term *word* is used for the group of bits constituting some information. Thus an 8-bit word might be the binary number 00100110. Each of the bits is communicated simultaneously along its own parallel wire. The system has four buses:

1 The *data bus* carries the data used in the processing carried out by the CPU. A microprocessor termed as being 8-bit has an internal data bus

which can handle 8-bit numbers. It can thus perform operations between 8-bit numbers and deliver results as 8-bit values.

2   The *address bus* is used to carry the addresses of memory locations. So that each word can be located in the memory, every memory location is given a unique *address*. Just like houses in a town are each given a distinct address so that they can be located, so each word location is given an address so that data stored at a particular location can be accessed by the CPU either to read data located there or put, i.e. write, data there. It is the address bus which carries the information indicating which address is to be accessed. If the address bus consists of 8 lines, the number of 8-bit words, and hence number of distinct addresses, is $2^8 = 256$. With 16 address lines, 65 536 addresses are possible.

3   The *control bus* carries the signals used by the CPU for control, e.g. to inform memory devices whether they are to receive data from an input or output data and to carry timing signals used to synchronise actions.

4   The *system bus* is used for communications between the input/output ports and the input/output unit

1.3.3 Memory

There are several memory elements in a PLC system:

1   System *read-only-memory (ROM)* to give permanent storage for the operating system and fixed data used by the CPU.

2   *Random-access memory (RAM)* for the user's program.

3   *Random-access memory (RAM)* for data. This is where information is stored on the status of input and output devices and the values of timers and counters and other internal devices. The data RAM is sometimes referred to as a *data table* or *register table*. Part of this memory, i.e. a block of addresses, will be set aside for input and output addresses and the states of those inputs and outputs. Part will be set aside for preset data and part for storing counter values, timer values, etc.

4   Possibly, as a bolt-on extra module, *erasable and programmable read-only-memory (EPROM)* for ROMS that can be programmed and then the program made permanent.

The programs and data in RAM can be changed by the user. All PLCs will have some amount of RAM to store programs that have been developed by the user and program data. However, to prevent the loss of programs when the power supply is switched off, a battery is used in the PLC to maintain the RAM contents for a period of time. After a program has been developed in RAM it may be loaded into an EPROM memory chip, often a bolt-on module to the PLC, and so made permanent. In addition there are temporary *buffer* stores for the input/output channels.

The storage capacity of a memory unit is determined by the number of binary words that it can store. Thus, if a memory size is 256 words then it can store $256 \times 8 = 2048$ bits if 8-bit words are used and $256 \times 16 = 4096$ bits if 16-bit words are used. Memory sizes are often specified in terms of the number of storage locations available with 1K representing the number 2^{10}, i.e. 1024. Manufacturers supply memory chips with the storage locations grouped in groups of 1, 4 and 8 bits. A 4K × 1 memory has $4 \times 1 \times 1024$ bit locations. A 4K × 8 memory has $4 \times 8 \times 1024$ bit locations. The term *byte* is used for a word of length 8 bits. Thus the 4K × 8 memory can store 4096 bytes. With a 16-bit address bus we can have 2^{16} different addresses and so, with 8-bit words stored at each address, we can have $2^{16} \times 8$ storage locations and so use a memory of size $2^{16} \times 8/2^{10} = 64K \times 8$ which we might have in the form of four 16K × 8 bit memory chips

1.3.4 Input/output unit

The input/output unit provides the interface between the system and the outside world, allowing for connections to be made through input/output channels to input devices such as sensors and output devices such as motors and solenoids. It is also through the input/output unit that programs are entered from a program panel. Every input/output point has a unique address which can be used by the CPU.

The input/output channels provide isolation and signal conditioning functions so that sensors and actuators can often be directly connected to them without the need for other circuitry. Electrical isolation from the external world is usually by means of *optoisolators* (the term *optocoupler* is also often used). Figure 1.9 shows the principle of an optoisolator. When a digital pulse passes through the light-emitting diode, a pulse of infrared radiation is produced. This pulse is detected by the photo transistor and gives rise to a voltage in that circuit. The gap between the light-emitting diode and the photo transistor gives electrical isolation but the arrangement still allows for a digital pulse in one circuit to give rise to a digital pulse in another circuit. The digital signal that is generally compatible with the microprocessor in the PLC is 5 V d.c. However, signal conditioning in the input channel, with isolation, enables a wide range of input signals to be supplied to it. A range of inputs might be available with a larger PLC, e.g. 5 V, 24 V, 110 V and 240 V digital/discrete, i.e. on–off, signals (Figure 1.10). A small PLC is likely to have just one form of input, e.g. 24 V. Figure 1.11 shows the basic form a d.c. input channel might take.

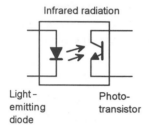

Figure 1.9 *Optoisolator*

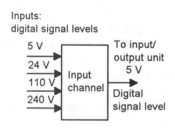

Figure 1.10 *Input levels*

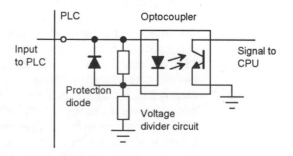

Figure 1.11 *Basic d.c. input circuit*

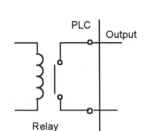

Figure 1.12 *Relay output*

Outputs are often specified as being of relay type, transistor type or triac type.

1 With the *relay type*, the signal from the PLC output is used to operate a relay and so is able to switch currents of the order of a few amperes in an external circuit. The relay not only allows small currents to switch much larger currents but also isolates the PLC from the external circuit. Relays are, however, relatively slow to operate. Relay outputs are suitable for a.c. and d.c. switching. They can withstand high surge currents and voltage transients. Figure 1.12 shows the basic feature of a relay output.

2 The *transistor type* of output uses a transistor to switch current through the external circuit. This gives a considerably faster switching action. It is, however, strictly for d.c. switching and is destroyed by overcurrent and high reverse voltage. As a protection, either a fuse or built-in electronic protection are used. Optoisolators are used to provide isolation. Figure 1.13 shows the basic form of such a transistor output channel.

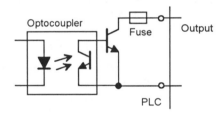

Figure 1.13 *Basic form of transistor output*

3 *Triac* outputs, with optoisolators for isolation, can be used to control external loads which are connected to the a.c. power supply. It is strictly for a.c. operation and is very easily destroyed by overcurrent. Fuses are virtually always included to protect such outputs.

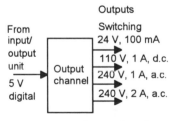

Figure 1.14 *Output levels*

The output from the input/output unit will be digital with a level of 5 V. However, after signal conditioning with relays, transistors or triacs, the output from the output channel might be a 24 V, 100 mA switching signal, a d.c. voltage of 110 V, 1 A or perhaps 240 V, 1 A a.c. or 240 V, 2 A a.c. from a triac output channel (Figure 1.14). With a small PLC, all the outputs might be of one type, e.g. 240 V a.c., 1 A. With modular PLCs, however, a range of outputs can be accommodated by selection of the modules to be used.

The following illustrates the types of inputs and outputs available with a small PLC, one of the Mitsubishi F2 series:

Number of inputs 12
Number of outputs 8
Input specification:

voltage 24 V ± 4 V, d.c.
operation current off → on, 4 mA d.c. max.,
on → on, 1.5 mA d.c. max.
Output specification:
Type: Relay
Relay isolation
2 A per point resistive load, 35 VA inductive load, 100 W lamp load
Type: Transistor
Optocoupler isolation
1 A per point resistive load, 24 W inductive load, 100 W lamp load
Type: Triac
Optocoupler isolation
1 A per point, 50 VA 110/120 V a.c., 100 VA 220/240 V a.c. inductive load, 100 W lamp load

Problems

Questions 1 to 6 have four answer options: A, B, C or D. Choose the correct answer from the answer options.

1 The term PLC stands for:

A Personal logic computer.
B Programmable local computer.
C Personal logic controller.
D Programmable logic controller.

2 Decide whether each of these statements is True (T) or False (F).

A transistor output channel from a PLC:
(i) Is used for only d.c. switching.
(ii) Is isolated from the output load by an optocoupler.
Which option BEST describes the two statements?

A (i) T (ii) T
B (i) T (ii) F
C (i) F (ii) T
D (i) F (ii) F

3 Decide whether each of these statements is True (T) or False (F).

A relay output channel from a PLC:
(i) Is used for only d.c. switching.
(ii) Can withstand transient overloads.
Which option BEST describes the two statements?

A (i) T (ii) T
B (i) T (ii) F
C (i) F (ii) T
D (i) F (ii) F

4   Decide whether each of these statements is True (T) or False (F).

A triac output channel from a PLC:
(i) Is used for only a.c. output loads.
(ii) Is isolated from the output load by an optocoupler.
Which option BEST describes the two statements?

A (i) T (ii) T
B (i) T (ii) F
C (i) F (ii) T
D (i) F (ii) F

5   Which of the following is most likely to be the voltage level used internally in a PLC, excluding the voltage levels that might occur during conditioning in output/input channels:

A 5 V
B 24 V
C 110 V
D 240 V

6   Decide whether each of these statements is True (T) or False (F).

The reason for including optocouplers on input/output units is to:
(i) Provide a fuse mechanism which breaks the circuit if high voltages or currents occur.
(ii) Isolate the CPU from high voltages or currents.
Which option BEST describes the two statements?

A (i) T (ii) T
B (i) T (ii) F
C (i) F (ii) T
D (i) F (ii) F

7   Draw a block diagram showing in very general terms the main units in a PLC.

8   Draw a block diagram of a PLC showing the main functional items and how buses link them, explaining the functions of each block.

9   State the characteristics of the relay, transistor and triac types of PLC output channels.

10   How many bits can a 2K memory store?

2 Input–output devices

This chapter is a brief consideration of typical input and output devices used with PLCs. The input devices considered include digital and analogue devices such as mechanical switches for position detection, proximity switches, photoelectric switches, encoders, temperature and pressure switches, potentiometers, linear variable differential transformers, strain gauges, thermistors, thermotransistors and thermocouples. Output devices considered include relays, contactors, solenoid valves and motors.

2.1 Input devices

Sensors which give digital/discrete, i.e. on–off, outputs can be easily connected to the input ports of PLCs. Sensors which give analogue signals have to be converted to digital signals before inputting them to PLC ports. The following are examples of some of the commonly used sensors.

2.1.1 Mechanical switches

A mechanical switch generates an on–off signal or signals as a result of some mechanical input causing the switch to open or close. Such a switch might be used to indicate the presence of a workpiece on a machining table, the workpiece pressing against the switch and so closing it. The absence of the workpiece is indicated by the switch being open and its presence by it being closed. Thus, with the arrangement shown in Figure 2.1(a), the input signals to a single input channel of the PLC are thus the logic levels:

Workpiece not present 0
Workpiece present 1

The 1 level might correspond to a 24 V d.c. input, the 0 to a 0 V input.

With the arrangement shown in Figure 2.1(b), when the switch is open the supply voltage is applied to the PLC input, when the switch is closed the input voltage drops to a low value. The logic levels are thus:

Workpiece not present 1
Workpiece present 0

Switches are available with *normally open (NO)* or *normally closed (NC)* contacts or can be configured as either by choice of the relevant contacts. An NO switch has its contacts open in the absence of a mechanical input and the mechanical input is used to close the switch. An NC switch has its contacts closed in the absence of a mechanical input and the mechanical input is used to open the switch.

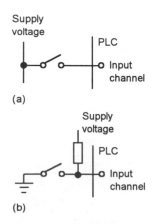

Figure 2.1 *Switch sensors*

The term *limit switch* is used for a switch which is used to detect the presence or passage of a moving part. It can be actuated by a cam, roller or lever. Figure 2.2 shows some examples. The cam (Figure 2.2(c)) can be rotated at a constant rate and so switch the switch on and off for particular time intervals.

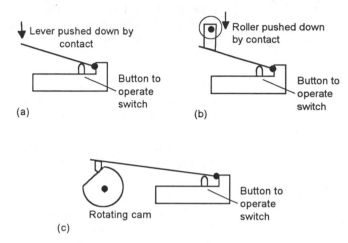

Figure 2.2 *Limit switches actuated by: (a) lever, (b) roller, (c) cam*

2.1.2 Proximity switches

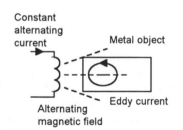

Figure 2.3 *Eddy current proximity switch*

Proximity switches are used to detect the presence of an item without making contact with it. There are a number of forms of such switches, some being only suitable for metallic objects.

The *eddy current* type of proximity switch has a coil which is energised by a constant alternating current and produces a constant alternating magnetic field. When a metallic object is close to it, eddy currents are induced in it (Figure 2.3). The magnetic field due to these eddy currents induces an e.m.f. back in the coil with the result that the voltage amplitude needed to maintain the constant coil current changes. The voltage amplitude is thus a measure of the proximity of metallic objects. The voltage can be used to activate an electronic switch circuit, basically a transistor which has its output switched from low to high by the voltage change, and so give an on–off device. The range over which such objects can be detected is typically about 0.5 to 20 mm.

Another type, the *inductive proximity switch*, consists of a coil wound round a ferrous metallic core. When one end of this core is placed near to a ferrous metal object there is effectively a change in the amount of metallic core associated with the coil and so a change in its inductance. This change in inductance can be monitored using a resonant circuit, the presence of the ferrous metal object thus changing the current in that circuit. The current can be used to activate an electronic switch circuit and so give an on–off device. The range over which such objects can be detected is typically about 2 to 15 mm.

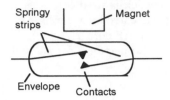

Figure 2.4 *Reed switch*

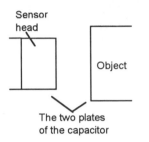

Figure 2.5 *Capacitive proximity switch*

Another type is the *reed switch*. This consists of two overlapping, but not touching, strips of a springy ferromagnetic material sealed in a glass or plastic envelope (Figure 2.4). When a magnet or current carrying coil is brought close to the switch, the strips become magnetised and attract each other. The contacts then close. The magnet closes the contacts when it is typically about 1 mm from the switch. Such a switch is widely used with burglar alarms to detect when a door is opened; the magnet being in the door and the reed switch in the frame of the door. When the door opens the switch opens.

A proximity switch that can be used with metallic and non-metallic objects is the *capacitive proximity switch*. The capacitance of a pair of plates separated by some distance depends on the separation, the smaller the separation the higher the capacitance. The sensor of the capacitive proximity switch is just one of the plates of the capacitor, the other plate being the metal object whose proximity is to be detected (Figure 2.5). Thus the proximity of the object is detected by a change in capacitance. The sensor can also be used to detect non-metallic objects since the capacitance of a capacitor depends on the dielectric between its plates. In this case the plates are the sensor and the earth and the non-metallic object is the dielectric. The change in capacitance can be used to activate an electronic switch circuit and so give an on–off device. Capacitive proximity switches can be used to detect objects when they are typically between 4 and 60 mm from the sensor head.

2.1.3 Photoelectric sensors and switches

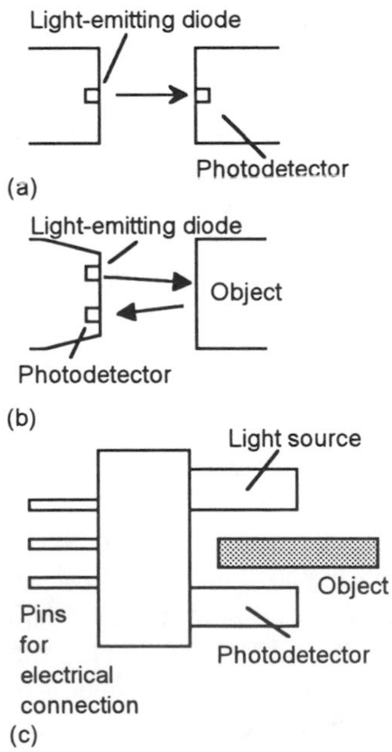

Photoelectric switch devices can either operate as *transmissive types* where the object being detected breaks a beam of light, usually infrared radiation, and stops it reaching the detector (Figure 2.6(a)) or *reflective types* where the object being detected reflects a beam of light onto the detector (Figure 2.6(b)). In both types the radiation emitter is generally a *light-emitting diode (LED)*. The radiation detector might be a *photo transistor*, often a pair of transistors, known as a *Darlington pair*. The Darlington pair increases the sensitivity. Depending on the circuit used, the output can be made to switch to either high or low when light strikes the transistor. Such sensors are supplied as packages for sensing the presence of objects at close range, typically at less than about 5 mm. Figure 2.6(c) shows a U-shaped form where the object breaks the light beam.

Another possibility is a *photo diode*. Depending on the circuit used, the output can be made to switch to either high or low when light strikes the diode. Yet another possibility is a *photo conductive cell*. The resistance of the photo conductive cell, often cadmium sulphide, depends on the intensity of the light falling on it.

With the above sensors, light is converted to a current, voltage or resistance change. If the output is to be used as a measure of the intensity of the light, rather than just the presence or absence of some object in the light path, the signal will need amplification and then conversion from analogue to digital by an analogue-to-digital converter. An alternative to

Figure 2.6 *Photoelectric sensors*

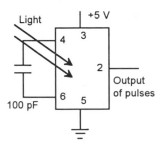

Figure 2.7 *TSL220*

this is to use a light-to-frequency converter, the light then being converted to a sequence of pulses with the frequency of the pulses being a measure of the light intensity. Integrated circuit sensors are available, e.g. the Texas Instrument TSL220, incorporating the light sensor and the voltage-to-frequency converter (Figure 2.7).

2.1.4 Encoders

The term *encoder* is used for a device that provides a digital output as a result of angular or linear displacement. An *increment encoder* detects changes in angular or linear displacement from some datum position, while an *absolute encoder* gives the actual angular or linear position.

Figure 2.8 shows the basic form of an incremental encoder for the measurement of angular displacement. A beam of light, from perhaps a light-emitting diode (LED), passes through slots in a disc and is detected by a light sensor, e.g. a photo diode or photo transistor. When the disc rotates, the light beam is alternately transmitted and stopped and so a pulsed output is produced from the light sensor. The number of pulses is proportional to the angle through which the disc has rotated, the resolution being proportional to the number of slots on a disc. With 60 slots then, since one revolution is a rotation of 360°, a movement from one slot to the next is a rotation of 6°. By using offset slots it is possible to have over a thousand slots for one revolution and so much higher resolution.

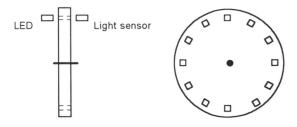

Figure 2.8 *Basic form of an incremental encoder*

The absolute encoder differs from the incremental encoder in having a pattern of slots which uniquely defines each angular position. Figure 2.9 shows the form of such an encoder using three sets of slots and so giving a 3-bit output. Typical encoders tend to have up to 10 or 12 tracks. The number of bits in the resulting binary output is equal to the number of tracks. Thus with 3 tracks there will be 3 bits and so the number of positions that can be detected is $2^3 = 8$, i.e. a resolution of 360/8 = 45°. With 10 tracks there will be 10 bits and the number of positions that can be detected is $2^{10} = 1024$ and the angular resolution is 360/1024 = 0.35°.

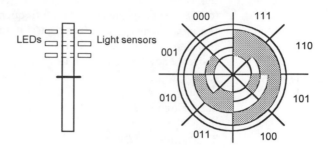

Figure 2.9 *A 3-bit absolute encoder*

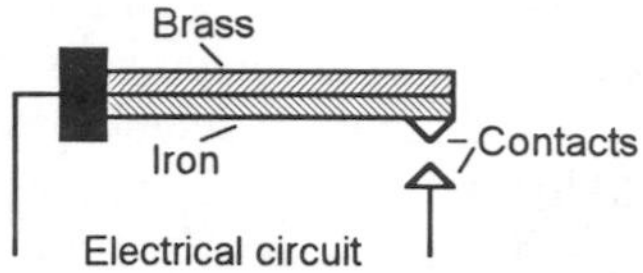

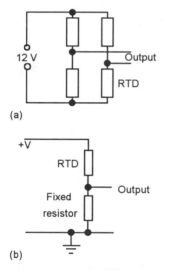

Figure 2.10 *Bimetallic strip*

Figure 2.11 *(a) Wheatstone bridge, (b) potential divider circuits*

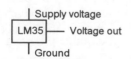

Figure 2.12 *LM35*

2.1.5 Temperature sensors

A simple form of temperature sensor which can be used to provide an on–off signal when a particular temperature is reached is the *bimetal element*. This consists of two strips of different metals, e.g. brass and iron, bonded together (Figure 2.10). The two metals have different coefficients of expansion. Thus when the temperature of the bimetal strip increases the strip curves, in order that one of the metals can expand more than the other. The higher expansion metal is on the outside of the curve. As the strip cools, the bending effect is reversed. This movement of the strip can be used to make or break electrical contacts and hence, at some particular temperature, give an on–off current in an electrical circuit. The device is not very accurate but is commonly used in domestic central heating thermostats.

Another form of temperature sensor is the *resistive temperature detector (RTD)*. The electrical resistance of metals or semiconductors changes with temperature. In the case of a metal, the ones most commonly used are platinum, nickel or nickel alloys, the resistance of which varies in a linear manner with temperature over a wide range of temperatures, though the actual change in resistance per degree is fairly small. Semiconductors, such as thermistors, show very large changes in resistance with temperature. The change, however, is non-linear. Such detectors can be used as one arm of a Wheatstone bridge and the output of the bridge taken as a measure of the temperature (Figure 2.11(a)). Another possibility is to use a potential divider circuit with the change in resistance of the thermistor changing the voltage drop across a resistor (Figure 2.11(b)). The output from either type of circuit is an analogue signal which is a measure of the temperature.

Thermodiodes and *thermotransistors* are used as temperature sensors since the rate at which electrons and holes diffuse across semiconductor junctions is affected by the temperature. Integrated circuits are available which combine such a temperature sensitive element with the relevant circuitry to give an output voltage related to temperature. A widely used integrated package is the LM35 which gives an output of 10 mV/°C when the supply voltage is +5 V (Figure 2.12).

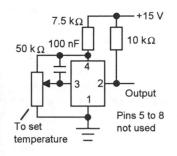

Figure 2.13 *LM3911N circuit for on–off control*

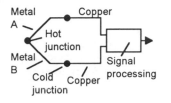

Figure 2.14 *Thermocouple*

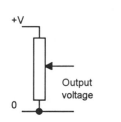

Figure 2.15 *Potentiometer*

A digital temperature switch can be produced with an analogue sensor by feeding the analogue output into a comparator amplifier which compares it with some set value, producing an output giving a logic 1 signal when the temperature voltage input is equal to or greater than the set point and otherwise an output which gives a logic 0 signal. Integrated circuits, e.g. LM3911N, are available combining a thermotransistor temperature sensitive element with an operational amplifier. When the connections to the chip are so made that the amplifier is connected as a comparator (Figure 2.13), then the output will switch as the temperature traverses the set point and so directly give an on–off temperature controller.

Another commonly used temperature sensor is the *thermocouple*. The thermocouple consists essentially of two dissimilar wires A and B forming a junction (Figure 2.14). When the junction is heated so that it is at a higher temperature than the other junctions in the circuit, which remain at a constant cold temperature, an e.m.f. is produced which is related to the hot junction temperature. The voltage produced by a thermocouple is small and needs amplification before it can be fed to the analogue channel input of a PLC. There is also circuitry required to compensate for the temperature of the cold junction since its temperature affects the value of the e.m.f. given by the hot junction. The amplification and compensation, together with filters to reduce the effect of interference from the 50 Hz mains supply, are often combined in a signal processing unit.

2.1.6 Displacement sensors

A *linear* or *rotary potentiometer* can be used to provide a voltage signal related to the position of the sliding contact between the ends of the potentiometer resistance track (Figure 2.15). The potentiometer thus provides an analogue linear or angular position sensor.

Another form is displacement sensor is the *linear variable differential transformer (LVDT)*, this giving a voltage output related to the position of a ferrous rod. The LVDT consists of three symmetrically placed coils through which the ferrous rod moves (Figure 2.16).

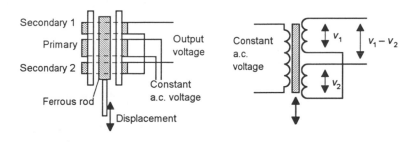

Figure 2.15 *LVDT*

When an alternating current is applied to the primary coil, alternating voltages are induced in the two secondary coils. When the ferrous rod core is centred between the two secondary coils, the voltages induced in them are equal. The outputs from the two secondary coils are connected so that their combined output is the difference between the two voltages. With the rod central, the two alternating voltages are equal and so there is no output voltage. When the rod is displaced from its central position there is more of the rod in one secondary coil than the other. As a result the size of the alternating voltage induced in one coil is greater than that in the other. The difference between the two secondary coil voltages, i.e. the output, thus depends on the position of the ferrous rod. The output from the LVDT is an alternating voltage. This is usually converted to an analogue d.c. voltage and amplified before inputting to the analogue channel of a PLC.

2.1.7 Strain gauges

When a wire or strip of semiconductor is stretched, its resistance changes. The fractional change in resistance is proportional to the fractional change in length, i.e. strain.

$$\frac{\Delta R}{R} = G \times \text{strain}$$

Figure 2.16 *Metal foil strain gauges*

where ΔR is the change in resistance for a wire of resistance R and G is a constant called the *gauge factor*. For metals the gauge factor is about 2 and for semiconductors about 100. Metal resistance strain gauges are in the form of a flat coil in order to get a reasonable length of metal in a small area. Often they are etched from metal foil (Figure 2.16) and attached to a backing of thin plastic film so that they can be stuck on surfaces, like postage stamps on an envelope. The change in resistance of the strain gauge, when subject to strain, is usually converted into a voltage signal by the use of a Wheatstone bridge (Figure 2.17).

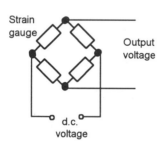

Figure 2.17 *Wheatstone bridge circuit*

A problem that occurs is that the resistance of the strain gauge also changes with temperature and thus some means of temperature compensation has to be used so that the output of the bridge is only a function of the strain. This can be achieved by placing a dummy strain gauge in an opposite arm of the bridge, that gauge not being subject to any strain but only the temperature (Figure 2.18). An alternative which is widely used is to use four active gauges as the arms of the bridge and arrange it so that one pair of opposite gauges are in tension and the other pair in compression. This not only gives temperature compensation but also gives a much larger output change when strain is applied. The following paragraph illustrates systems employing such a form of compensation.

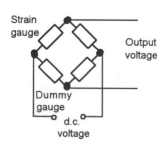

Figure 2.18 *Temperature compensation*

By attaching strain gauges to other devices, changes which result in strain of those devices can be transformed, by the strain gauges, to give voltage changes. They might, for example, be attached to a cantilever to which forces are applied at its free end (Figure 2.19(a)). The voltage change, resulting from the strain gauges and the Wheatstone bridge, then becomes a measure of the force. Another possibility is to attach strain gauges to a diaphragm which deforms as a result of pressure (Figure 2.19(b)). The output from the gauges, and associated Wheatstone bridge, then becomes a measure of the pressure.

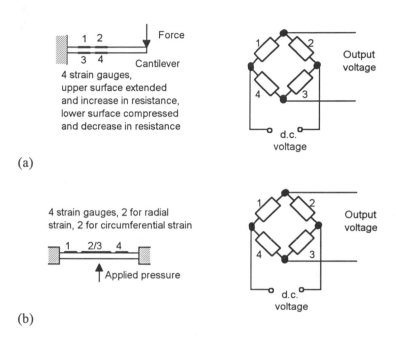

(a)

(b)

Figure 2.19 *Strain gauges used for (a) force sensor, (b) pressure sensor*

2.1.8 Pressure sensors

Commonly used pressure sensors which gives responses related to the pressure are diaphragm and bellows types. The diaphragm type consists of a thin disk of metal or plastic, secured round its edges. When there is a pressure difference between the two sides of the diaphragm, the centre of it deflects. The amount of deflection is related to the pressure difference. This deflection may be detected by strain gauges attached to the diaphragm (see Figure 2.19(b)) or by using the deflection to squeeze a piezoelectric crystal (Figure 2.20). When a piezoelectric crystal is squeezed, there is a relative displacement of positive and negative charges within the crystal and the outer surfaces of the crystal become charged. Hence a potential difference appears across it. An example of such a sensor is the Motorola MPX100AP sensor (Figure 2.21). This has a built-in vacuum on one side of the diaphragm and so the deflection of the diaphragm gives a measure of the absolute pressure applied to the other side of the diaphragm. The output is a voltage which is proportional to the applied pressure with a sensitivity of 0.6 mV/kPa. Other versions are available which have one side of the diaphragm open to the atmosphere and so can be used to measure gauge

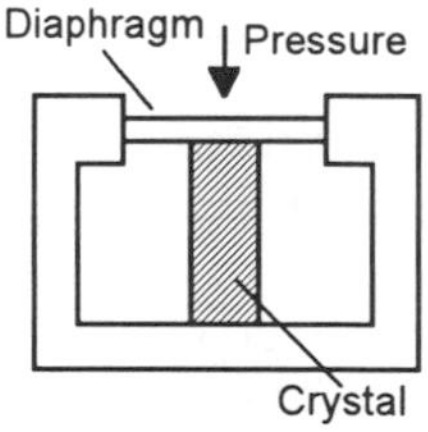

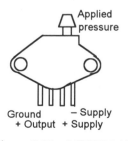

Figure 2.20 *Piezoelectric pressure sensor*

Figure 2.21 *MPX100AP*

pressure, others allow pressures to be applied to both sides of the diaphragm and so can be used to measure differential pressures.

Pressure switches are designed to switch on or off at a particular pressure. A typical form involves a diaphragm or bellows which moves under the action of the pressure and operates a mechanical switch. Figure 2.22 shows two possible forms. Diaphragms are less sensitive than bellows but can withstand greater pressures.

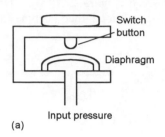

(a)

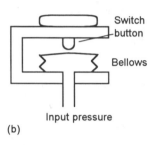

(b)

Figure 2.22 *Examples of pressure switches*

2.1.9 Liquid level detector

Pressure sensors may be used to monitor the depth of a liquid in a tank. The pressure due to a height of liquid h above some level is $h\rho g$, where ρ is the density of the liquid and g the acceleration due to gravity. Thus a commonly used method of determining the level of liquid in a tank is to measure the pressure due to the liquid above some datum level (Figure 2.23).

Often a sensor is just required to give a signal when the level in some container reaches a particular level. A float switch that is used for this purpose consists of a float containing a magnet which moves in a housing with a reed switch. As the float rises of falls it turns the reed switch on or off, the reed switch being connected in a circuit which then switches on or off a voltage.

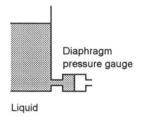

Figure 2.23 *Liquid level sensor*

2.1.10 Fluid flow measurement

A common form of fluid flow meter is that based on measuring the difference in pressure resulting when a fluid flows through a constriction. Figure 2.24 shows a commonly used form, the *orifice flow meter*. As a result of the fluid flowing through the orifice, the pressure at A is higher than that at B, the difference in pressure being a measure of the rate of flow. This pressure difference can be monitored by means of a diaphragm pressure gauge and thus becomes a measure of the rate of flow.

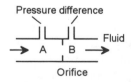

Figure 2.24 *Orifice flow meter*

2.1.11 Keypads

Many machines employ keypads to input instructions to set the conditions required for outputs such as temperatures or speeds. Such keypads commonly have buttons which, when pressed, operate conductive silicon rubber pads to make contacts. Figure 2.25 shows the form such a 12-way keypad can take. Rather than have each key wired up separately and so giving 12 inputs, the keys are connected in rows and columns and closing a single key can give a column output and a row output which is unique to that key. This reduces the number of inputs required to the PLC.

Figure 2.25 *12-way keypad*

2.2 Output devices

The output ports of a PLC are of the relay type or optoisolator with transistor or triac types depending on the devices connected to them which are to be switched on or off (see Section 1.3.4). Generally, the digital signal from an output channel of a PLC is used to control an actuator which in turn controls some process. The term *actuator* is used for the device which transforms the electrical signal into some more powerful action which then results in the control of the process. The following are some examples.

2.2.1 Contactor

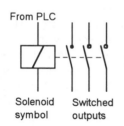

Solenoid   Switched
symbol   outputs

Figure 2.26 *Contactor*

Solenoids form the basis of a number of output control actuators. When a current passes through a solenoid a magnetic field is produced and this can then attract ferrous metal components in its vicinity. One example of such an actuator is the *contactor*. When the output from the PLC is switched on, the solenoid magnetic field is produced and pulls on the contacts and so closes a switch or switches (Figure 2.26). The result is that much larger currents can be switched on. Thus the contactor might be used to switch on the current to a motor.

Essentially a contactor is a form of *relay*, the difference being that the term relay is used for a device for switching small currents, less than about 10 A, whereas the term contactor is used for a heavy current switching device with currents up to many hundreds of amps.

2.2.2 Directional control valves

Another example of the use of a solenoid as an actuator is a *solenoid operated valve*. The valve may be used to control the directions of flow of pressurised air or oil and so used to operate other devices such as a piston moving in a cylinder. Figure 2.27 shows one such form, a spool valve, used to control the movement of a piston in a cylinder.

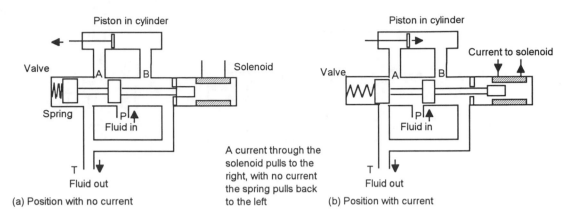

Figure 2.27 *An example of a solenoid operated valve*

Figure 2.28 *Two position valve*

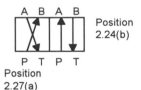

Figure 2.29 *The 4/2 valve*

2/2 valve: flow from P to A switched to no flow

3/2 valve: no flow from P to A and flow from A to T switched to T being closed and flow from P to A

Figure 2.30 *Direction valves*

Pressurised air or hydraulic fluid is inputted from port P, this being connected to the pressure supply from a pump or compressor and port T is connected to allow hydraulic fluid to return to the supply tank or, in the case of a pneumatic system, to vent the air to the atmosphere. With no current through the solenoid (Figure 2.27(a)) the hydraulic fluid of pressurised air is fed to the right of the piston and exhausted from the left, the result then being the movement of the piston to the left. When a current is passed through the solenoid, the spool valve switches the hydraulic fluid or pressurised air to the left of the piston and exhausted from the right. The piston them moves to the right. The movement of the piston might be used to push a deflector to deflect items off a conveyor belt (see Figure 1.1(b)) or implement some other form of displacement which requires power.

With the above valve there are the two control positions shown in Figure 2.27(a) and (b). Directional control valves are described by the number of ports they have and the number of control positions. The valve shown in Figure 2.27 has four ports, i.e. A, B, P and T, and two control positions. It is thus referred to as a 4/2 valve. The basic symbol used on drawings for valves is a square, with one square being used to describe each of the control positions. Thus the symbol for the valve in Figure 2.27 consists of two squares (Figure 2.28). Within each square the switching positions are then described by arrows to indicate a flow direction or a terminated line to indicate no flow path. Figure 2.29 shows this for the valve shown in Figure 2.27. Figure 2.30 shows some more examples of direction valves and their switching positions.

The actuation methods used with valves are added to the diagram symbol; Figure 2.31 shows examples of such symbols. The value shown in Figure 2.27 has a spring to give one position and a solenoid to give the other and so the symbol is as shown in Figure 2.32.

Direction valves can be used to control the direction of motion of pistons in cylinders, the displacement of the pistons being used to implement the required actions. The term *single acting cylinder* (Figure 2.33(a)) is used

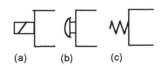

Figure 2.31 *Actuation: (a) solenoid, (b) push button, (c) spring operated*

for one which is powered by the pressurised fluid being applied to one side of the piston to give motion in one direction, it being returned in the other direction by possibly an internal spring. The term *double acting cylinder* (Figure 2.33(b)) is used when the cylinder is powered by fluid for its motion in both piston movement directions. Figure 2.34 shows how a valve can be used to control the direction of motion of a piston in a single-acting cylinder; Figure 2.35 shows how two valves can be used to control the action of a piston in a double acting cylinder.

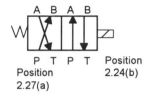

Figure 2.32 *The 4/2 valve*

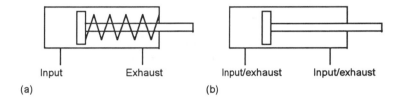

Figure 2.33 *Cylinders: (a) single acting, (b) double acting*

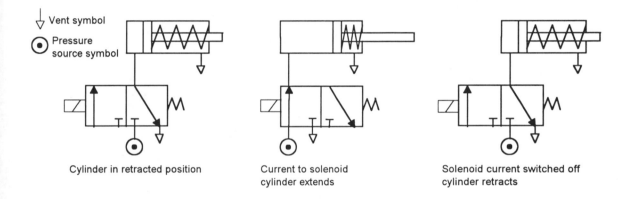

Figure 2.34 *Control of a single-acting cylinder*

2.2.2 Motors

A *d.c. motor* has coils of wire mounted in slots on a cylinder of ferromagnetic material, this being termed the *armature*. The armature is mounted on bearings and is free to rotate. It is mounted in the magnetic field produced by permanent magnets or current passing through coils of wire, these being termed the *field coils*. The permanent magnet or electromagnet is termed the *stator*. When a current passes through the armature coil, because a current carrying conductor with a magnetic field at right angles to it experiences a force, forces act on the coil and result in rotation. Figure 2.36 shows the basic principles of such a motor. Brushes and a commutator are used to reverse the current through the coil every half rotation and so keep the coil rotating.

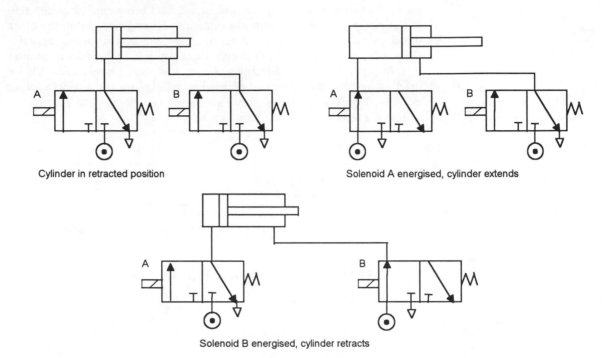

Figure 2.35 *Control of a double acting cylinder*

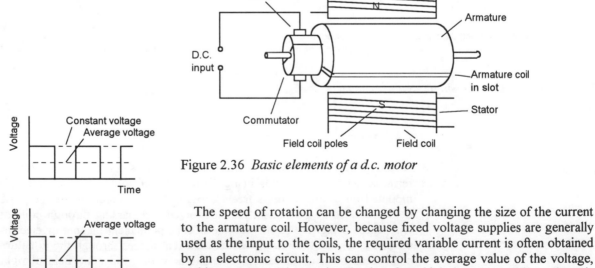

Figure 2.36 *Basic elements of a d.c. motor*

Figure 2.37 *Pulse width modulation*

The speed of rotation can be changed by changing the size of the current to the armature coil. However, because fixed voltage supplies are generally used as the input to the coils, the required variable current is often obtained by an electronic circuit. This can control the average value of the voltage, and hence current, by varying the time for which the constant d.c. voltage is switched on (Figure 2.37). The term *pulse width modulation (PWM)* is used since the width of the voltage pulses is used to control the average d.c. voltage applied to the armature. A PLC might thus control the speed of

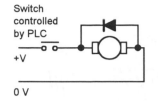

Figure 2.38 *On–off control*

rotation of a motor by controlling the electronic circuit used to control the width of the voltage pulses.

Many industrial processes only require the PLC to switch a d.c. motor on or off. This might be done using a contactor (see section 2.3.1). Figure 2.38 shows the basic principle. The diode is included to dissipate the induced current resulting from the back e.m.f. Sometimes a PLC is required to reverse the direction of rotation of the motor. This can be done using relays or contactors to reverse the direction of the current applied to the armature coil. Figure 2.39 shows the basic principle. For rotation in one direction, switch 1 is closed and switch 2 opened. For rotation in the other direction, switch 1 is opened and switch 2 closed

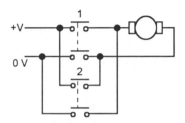

Figure 2.39 *Direction control for a d.c. motor*

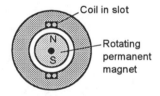

Figure 2.40 *Principle of brushless d.c. motor*

Another form of d.c. motor is the *brushless d.c. motor*. This uses a permanent magnet for the magnetic field but, instead of the armature coil rotating as a result of the magnetic field of the magnet, the permanent magnet rotates within the stationary coil. Figure 2.40 shows the basic principle, just one coil being shown. With the conventional d.c. motor, a commutator has to be used to reverse the current through the coil every half rotation in order to keep the coil rotating in the same direction. With the brushless permanent magnet motor, electronic circuitry is used to reverse the current.

The motor can be started and stopped by controlling the current to the stationary coil. To reverse the motor, reversing the current is not so easy because of the electronic circuitry used for the commutator function. One method that is used is to incorporate sensors with the motor to detect the position of the north and south poles. These sensors can then cause the current to the coils to be switched at just the right moment to reverse the forces applied to the magnet. The speed of rotation can be controlled using pulse width modulation, i.e. controlling the average value of pulses of a constant d.c. voltage (see Figure 2.37).

Alternating current motors consist of two basic parts, a rotating cylinder called the *rotor* and a stationary part called the *stator*. The stator surrounds the rotor and has the coil windings that produce a rotating magnetic field in the space occupied by the rotor. It is this rotating magnetic field which causes the rotor to rotate. One form of such a motor is illustrated in Figure 2.41. This is the *single-phase squirrel-cage induction motor*.

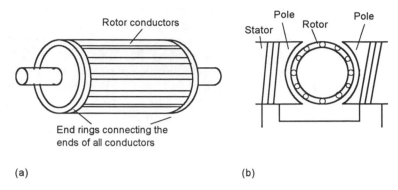

Rotor conductors

End rings connecting the
ends of all conductors

Pole   Pole
Stator   Rotor

(a)   (b)

Figure 2.41   *(a) Squirrel-cage rotor, (b) the rotor with a single-phase stator*

The rotor is the squirrel cage, consisting of copper or aluminium bars fitting into slots in end rings to form a set of parallel connected conductors. There are no external electrical connections to the rotor. When an alternating current passes through the stator coil, an alternating magnetic field is produced and, as a consequence, an e.m.f. is induced in the rotor conductors and currents flow through them. We thus have currents flowing in conductors in a magnetic field and so forces act on them. Given an initial impetus, these forces continue the rotation. The rotor rotates at a speed determined by the frequency of the alternating current applied to the stator. One way of varying the speed of rotation is to use an electronic circuit to control the frequency of the current supplied to the stator.

Though a.c. motors are cheaper, more rugged and more reliable than d.c. motors, the maintaining of constant speed and controlling that speed is generally more complex than with d.c. motors. As a consequence, d.c. motors, particularly brushless permanent magnet motors, tend to be more widely used for control purposes.

2.2.3 Stepper motors

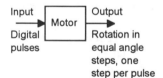

Input   Output
→   Motor   →
Digital   Rotation in
pulses   equal angle
               steps, one
               step per pulse

Figure 2.42   *The stepping motor*

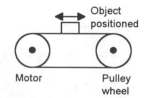

Object
positioned

Motor   Pulley
               wheel

Figure 2.43   *Linear positioning*

The *stepper* or *stepping motor* is a motor that produces rotation through equal angles, the so-termed *steps*, for each digital pulse supplied to its input (Figure 2.42). Thus, if one input pulse produces a rotation of 1.8° then 20 such pulses would give a rotation of 36.0°. To obtain one complete revolution through 360°, 200 digital pulses would be required. The motor can thus be used for accurate angular positioning. If it is used to drive a continuous belt (Figure 2.43), it can be used to give accurate linear positioning. Such a motor is used with computer printers, robots, machine tools and a wide range of instruments where accurate positioning is required.

There are a number of forms of stepping motor. Figure 2.44 shows the basic principle of the *variable reluctance* type. The rotor is made of soft steel and has a number of teeth, the number being less than the number of poles on the stator. The stator has pairs of poles, each pair of poles being activated and made into an electromagnet by a current being passed through the coils wrapped round them. When one pair of poles is activated, a magnetic field is produced which attracts the nearest pair of rotor teeth so that the teeth and poles line up. This is termed the position of *minimum reluctance*. By then switching the current to the next pair of poles, the rotor can be made to rotate to line up with those poles. Thus by sequentially switching the current from one pair of poles to the next, the rotor can be made to rotate in steps.

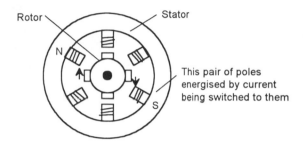

Figure 2.44 *The principle of the variable reluctance stepper motor*

To drive a stepper motor, each pair of stator coils has to be switched on and off in the required sequence. Thus the input to the motor of a sequence of pulses has to provide outputs to each of the pairs of stator coils in the correct sequence. The drive system used for this purpose consists essentially of two blocks, a logic sequencer and a driver (Figure 2.45(a)). The logic sequencer takes the input of the pulses and gives the required sequence of outputs to control the driver so that it produces the required size outputs, in sequence, to activate the coils of the stepper motor. Figure 2.45(b) illustrates such a sequence for a stator having four pairs of coils.

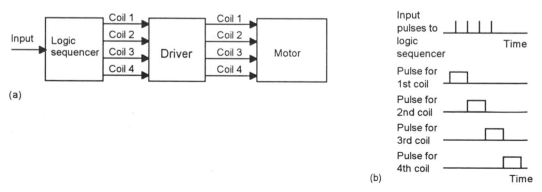

Figure 2.45 *(a) Drive system for a four-phase stepper motor, (b) input and outputs of the drive system*

Driver circuits can be obtained as integrated circuits. Figure 2.46 shows the integrated circuit SAA1027 and its connections for use with a stepper motor having four pairs of stator poles. The input to trigger the step rotation of the stepper motor is a low to high transition for the voltage on the input to pin 15. Each such trigger results in a rotation of one step. The outputs from the integrated circuit are currents in sequence along the brown (pin 6), black (pin 8), green (pin 9) and yellow (pin 11) connections to the stator coils. The motor will run clockwise when pin 3 is low, i.e. less that 4.5 V, and anticlockwise when it is high, i.e. more than 7.5 V. When pin 2 is made low, the output resets to its initial position.

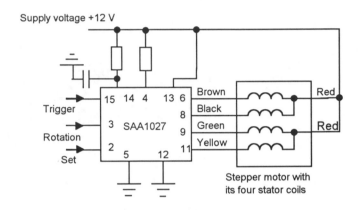

Figure 2.46 *Driver circuit connections with the integrated circuit SAA1027*

2.3 Examples of applications

The following are some examples of control systems designed to illustrate the use of a range of input and output devices.

2.3.1 A conveyor belt

Consider a conveyor belt that is to be used to transport goods from a loading machine to a packaging area (Figure 2.47). When an item is loaded onto the conveyor belt, a contact switch might be used to indicate that the item is on the belt and start the conveyor motor. The motor then has to keep running until the item reaches the far end of the conveyor and falls off into the packaging area. When it does this, a switch might be activated which has the effect of switching off the conveyor motor. The motor is then to remain off until the next item is loaded onto the belt. Thus the inputs to a PLC controlling the conveyor are from two switches and the output is to a motor.

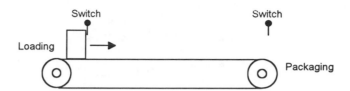

Figure 2.47 *Conveyor*

2.3.2 A lift

Consider a simple goods lift to move items from one level to another. It might be bricks from the ground level to the height where the bricklayers are working. The lift is to move upwards when a push button is pressed at the ground level to send the lift upwards or a push button is pressed at the upper level to request the lift to move upwards, but in both cases there is a condition that has to be met that a limit switch indicates that the access gate to the lift platform is closed. The lift is to move downwards when a push button is pressed at the upper level to send the lift downwards or a push button is pressed at the lower level to request the lift to move downwards, but in both cases there is a condition that has to be met that a limit switch indicates that the access gate to the lift platform is closed. Thus the inputs to the control system are electrical on-off signals from push button switches and limit switches. The output from the control system is the signal to control the motor.

2.3.3 An automatic door

Consider an automatic door which is to open when a person approaches it, remain open for a specified time, say 5 s, before closing. The inputs to the control system might be from a sensor to detect a person approaching from the outside and another sensor to detect a person approaching from the inside. These sensors might be heat sensitive semiconductor elements which give voltage signals when infrared radiation falls on them. There will also be inputs to the controller probably from limit switches to indicate when the door is fully open and a timer to keep the door open for the required time. The output from the controller might be to solenoid operated pneumatic valves which use the movement of pistons in a cylinder to open and close the door. Figure 2.48 shows a simple valve system that might be used.

When there is an output to the solenoid to open the door inwards, because a person has approached from the outside, the air pressure is applied, via port P, to the unvented side of the piston and causes it to move. When this solenoid is no longer energised, the spring returns the piston back by connecting the unvented side to a vent to the atmosphere. A similar arrangement is used for opening the door outwards.

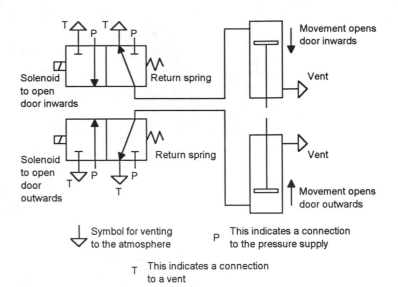

Figure 2.48 *Pneumatic door opening system*

Problems

Questions 1 to 10 have four answer options: A, B, C or D. Choose the correct answer from the answer options.

1 Decide whether each of these statements is True (T) or False (F).

A limit switch:
(i) Can be used to detect the presence of a moving part.
(ii) Is activated by contacts making or breaking an electrical circuit.

A (i) T (ii) T
B (i) T (ii) F
C (i) F (ii) T
D (i) F (ii) F

2 Decide whether each of these statements is True (T) or False (F).

A thermistor is a temperature sensor which gives resistance changes which are:
(i) A non-linear function of temperature.
(ii) Large for comparatively small temperature changes.

A (i) T (ii) T
B (i) T (ii) F
C (i) F (ii) T
D (i) F (ii) F

3 A diaphragm pressure sensor is required to give a measure of the gauge pressure present in a system. Such a sensor will need to have a diaphragm with:

A A vacuum on one side
B One side open to the atmosphere
C The pressure applied to both sides
D A controlled adjustable pressure applied to one side

4 The change in resistance of an electrical resistance strain gauge with a gauge factor of 2.0 and resistance 100 Ω when subject to a strain of 0.001 is:

A 0.0002 Ω
B 0.002 Ω
C 0.02 Ω
D 0.2 Ω

5 An incremental shaft encoder gives an output which is a direct measure of:

A The diameter of the shaft.
B The change in diameter of the shaft.
C The change in angular position of the shaft.
D The absolute angular position of the shaft.

6 Decide whether each of these statements is True (T) or False (F).

Input devices which can be used to give an analogue input for displacement are:
(i) Linear potentiometer.
(ii) Linear variable differential transformer.

A (i) T (ii) T
B (i) T (ii) F
C (i) F (ii) T
D (i) F (ii) F

Questions 7 and 8 refer to Figure 2.49 which shows the symbol for a directional valve.

7 Decide whether each of these statements is True (T) or False (F).

The valve has:
(i) 4 ports
(ii) 2 positions

A (i) T (ii) T
B (i) T (ii) F
C (i) F (ii) T
D (i) F (ii) F

Figure 2.49 *Problems 7 and 8*

8   Decide whether each of these statements is True (T) or False (F).

In the control positions:
(i)   A is connected to T and P to B.
(ii) P is connected to A and B to T.

A   (i) T   (ii) T
B   (i) T   (ii) F
C   (i) F   (ii) T
D   (i) F   (ii) F

9   Decide whether each of these statements is True (T) or False (F).

A stepper motor is specified as having a step angle of 1.8°. This means that:
(i)   Each pulse input to the motor rotates the motor shaft by 1.8°.
(ii) The motor shaft takes 1 s to rotate through 1.8°.

A   (i) T   (ii) T
B   (i) T   (ii) F
C   (i) F   (ii) T
D   (i) F   (ii) F

10 Decide whether each of these statements is True (T) or False (F).

A proximity switch is required for detecting the presence of a non-metallic object. Types of switches that might be suitable are:
(i)   Eddy current type.
(ii) Capacitive type.

A   (i) T   (ii) T
B   (i) T   (ii) F
C   (i) F   (ii) T
D   (i) F   (ii) F

11 Explain the operation of the following input devices, stating the form of the signal being sensed and the output:
(a) Reed switch.
(b) Incremental shaft encoder.
(c) Photoelectric transmissive switch.
(d) Diaphragm pressure switch.

12 Explain how the on–off operation and direction of a d.c. motor can be controlled by switches.

13 Explain the principle of the stepper motor.

3 Input/output processing

This chapter continues the discussion of inputs and outputs from Chapter 2 and is a brief consideration of the processing of the signals from input and output devices. It includes the forms of typical input/output modules and, in an installation where sensors are some distance from the PLC processing, their communication links to the PLC.

3.1 Input/output units

Input signals from sensors and outputs to actuating devices can be:

1 *Analogue*, i.e. a signal whose size is related to the size of the quantity being sensed.
2 *Discrete*, i.e. essentially just an on–off signal.
3 *Digital*, i.e. a sequence of pulses.

The CPU, however, must have an input of digital signals of a particular size, normally 0 to 5 V. The output from the CPU is digital, normally 0 to 5 V. Thus there is a need to manipulate input and output signals so that they are in the required form.

The input/output units of PLCs are designed so that a range of input signals can be changed into 5 V digital signals and so that a range of outputs are available to drive external devices. It is this in-built facility to enable a range of inputs and outputs to be handled which makes PLCs so easy to use. In general, the input range on a particular input module is selected by means of DIP switches. DIP is short for Dual-In-Line Package. These switches are located in the back of the module (Figure 3.1). Such switches are either on or off and are used to set the parameters for a particular module. The switches are also used to set addresses for modules.

The following is a brief indication of the basic circuits used for input and output units. In the case of rack instruments they are mounted on cards which can be plugged into the racks. The input/output characteristics of a PLC can thus be changed by changing the cards. A single box form of PLC has input/output units incorporated by the manufacturer.

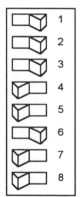

Figure 3.1   *DIP switches*

3.1.1 Input units

Figures 3.2 and 3.3 shows the basic input unit circuits for d.c. and a.c. discrete and digital inputs. Optoisolators (see Section 1.3.4) are used to provide protection. With the a.c. input card a rectifier bridge network is used to rectify the a.c. so that the resulting d.c. signal can provide the signal for use by the optoisolator to give the input signals to the CPU of the PLC.

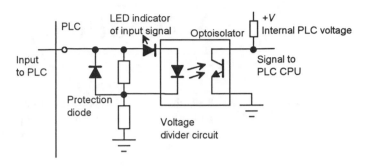

Figure 3.2 *D.C. input unit*

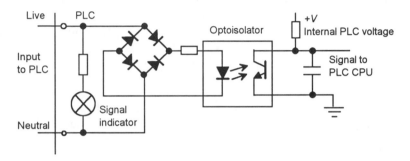

Figure 3.3 *A.C. input unit*

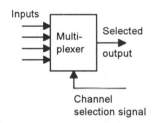

Figure 3.4 *Multiplexer*

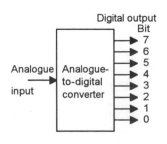

Figure 3.5 *Function of an analogue-to-digital converter*

Analogue signals can be inputted to a PLC if the input channel is able to convert the signal to a digital signal using an analogue-to-digital converter. With a rack mounted system this may be achieved by mounting a suitable analogue input card in the rack. So that one analogue card is not required for each analogue input, multiplexing is generally used (Figure 3.4). This involves more than one analogue input being connected to the card and then electronic switches used to select each input in turn. Cards are typically available giving 4, 8 or 16 analogue inputs.

Figure 3.5 illustrates the function of an analogue-to-digital converter (ADC). A single analogue input signal gives rise to on–off signals along perhaps eight separate wires. The eight signals then constitute the so-termed digital *word* corresponding to the analogue input signal level. With such an 8-bit converter there are $2^8 = 256$ different digital values possible; these are 0000 0000 to 1111 1111, i.e. 0 to 255. The digital output goes up in steps (Figure 3.6) and the analogue voltages required to produce each digital output are termed *quantisation levels*. The analogue voltage has to change by the difference in analogue voltage between successive levels if the binary output is to change. The term *resolution* is used for the smallest change in analogue voltage which will give rise to a change in one bit in the digital output. With an 8-bit ADC, if, say, the

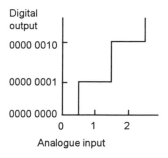

Figure 3.6 *Steps*

full-scale analogue input signal varies between 0 and 10 V then a step of one digital bit involves an analogue input change of 10/255 V or about 0.04 V. This means that a change of 0.03 V in the input will produce no change in the digital output. The number of bits in the output from an analogue-to-digital converter thus determines the *resolution*, and hence *accuracy*, possible. If a 10-bit ADC is used then $2^{10} = 1024$ different digital values are possible and, for the full-scale analogue input of 0 to 10 V, a step of one digital bit involves an analogue input change of 10/1023 V or about 0.01 V. If a 12-bit ADC is used then $2^{12} = 4096$ different digital values are possible and, for the full-scale analogue input of 0 to 10 V, a step of one digital bit involves an analogue input change of 10/4095 V or about 2.4 mV. In general, the resolution of an *n*-bit ADC is $1/(2^n - 1)$.

The following illustrates the analogue-to-digital conversion for an 8-bit converter when the analogue input is in the range 0 to 10 V:

Analogue input (V)	Digital output (V)
0.00	0000 0000
0.04	0000 0001
0.08	0000 0010
0.12	0000 0011
0.16	0000 0100
0.20	0000 0101
0.24	0000 0110
0.28	0000 0111
0.32	0000 1000
etc.	

To illustrate the above, consider a thermocouple used as a sensor with a PLC and giving an output of 0.5 mV per °C. What will be the accuracy with which the PLC will activate the output device if the thermocouple is connected to an analogue input with a range of 0 to 10 V d.c and using a 10-bit analogue-to-digital converter? With a 10-bit converter there is $2^{10} = 1024$ bits covering the 0 to 10 V range. Thus a change of 1 bit corresponds to 10/1023 V or about 0.01 V, i.e. 10 mV. Hence the accuracy with which the PLC recognises the input from the thermocouple is ±5 mV or ±10°C.

3.1.2 Output units

Output units can be relay, transistor or triac. Figure 3.6 shows the basic form of a relay output unit, Figure 3.7 that of a transistor output unit and Figure 3.8 that of a triac output unit.

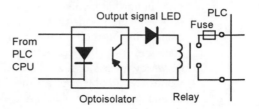

Figure 3.6 *Relay output unit*

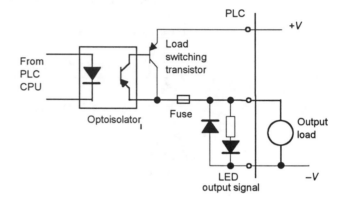

Figure 3.7 *Transistor output unit*

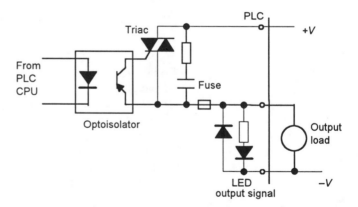

Figure 3.8 *Triac output unit*

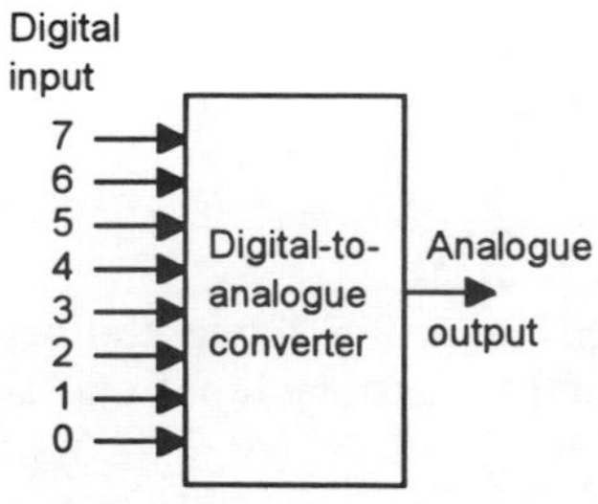

Digital
input

Figure 3.10 *DAC function*

Analogue outputs are frequently required and can be provided by digital-to-analogue converters (DACs) at the output channel. The input to the converter is a sequence of bits with each bit along a parallel line. Figure 3.9 shows the basic function of the converter. A bit in the 0 line gives rise to a certain size output pulse. A bit in the 1 line gives rise to an output pulse of twice the size of the 0 line pulse. A bit in the 2 line gives rise to an output

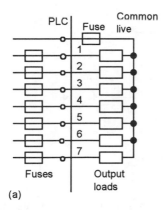

(a)

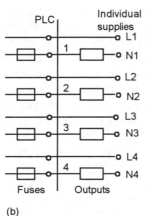

(b)

Figure 3.11 *Forms of output:*
(a) common supply,
(b) individual supplies

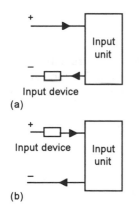

Figure 3.12 *Input unit:*
(a) sourcing, (b) sinking

pulse of twice the size of the 1 line pulse. A bit in the 3 line gives rise to an output pulse of twice the size of the 2 line pulse, and so on. All the outputs add together to give the analogue version of the digital input. When the digital input changes, the analogue output changes in a stepped manner, the voltage changing by the voltage changes associated with each bit. For example, if we have an 8-bit converter then the output is made up of voltage values of $2^8 = 256$ analogue steps. Suppose the output range is set to 10 V d.c. One bit then gives a change of 10/255 V or about 0.04 V. Thus we have:

Digital input (V)	Analogue output (V)
00000000	0.00
00000001	0.04
00000010	0.08 + 0.00 = 0.08
00000011	0.08 + 0.04 = 0.12
00000100	0.16
00000101	0.016 + 0.00 + 0.04 = 0.20
00000110	0.016 + 0.08 = 0.24
00000111	0.016 + 0.08 + 0.04 = 0.28
00001000	0.32
etc.	

Analogue output modules are usually provided in a number of outputs, e.g. 4 to 20 mA, 0 to +5 V d.c., 0 to +10 V d.c., and the appropriate output is selected by DIP switches on the module. Modules generally have outputs in two forms, one for which all the outputs from that module have a common voltage supply and one which drives outputs having their own individual voltage supplies. Figure 3.11 shows the basic principles of these two forms of output.

3.1.3 Sourcing and sinking

The terms *sourcing* and *sinking* refer to the manner in which d.c. devices are interfaced with the PLC. For a PLC input unit, with sourcing it is the source of the current supply for the input device connected to it (Figure 3.12 (a)). With sinking, the input device provides the current to the input unit (Figure 3.12(b)). With a PLC output unit, when it provides the current for the output device (Figure 3.13(a)) it is said to be sourcing and when the output device provides the current to the output unit it is said to be sinking (Figure 3.13(b)). Quite often, sinking output units are used for interfacing with electronic equipment and sourcing output units for interfacing with solenoids.

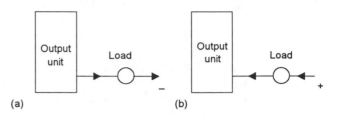

Figure 3.13 *Output unit: (a) sourcing, (b) sinking*

3.2 Signal conditioning

When connecting to an input unit, sensors which generate digital or discrete signals, care has to be taken to ensure that voltage levels match. However, many sensors generate analogue signals. In order to avoid having a multiplicity of analogue input channels to cope with the wide diversity of analogue signals that can be generated by sensors, external signal conditioning is often used to bring analogue signals to a common range and so allow a standard form of analogue input channel to be used. A common standard that is used (Figure 3.14) is to convert analogue signals to a current in the range 4 to 20 mA and thus to a voltage by passing it through a 250 Ω resistance to give a 1 to 5 V input signal. Thus, for example, a sensor used to monitor liquid level in the height range 0 to 1 m would have the 0 level represented by 4 mA and the 1 m represented by 20 mA. The use of 4 mA to represent the low end of the analogue range serves the purpose of distinguishing between when the sensor is indicating zero and when the sensor is not working and giving zero response for that reason. When this happens the current would be 0 mA. The 4 mA also is often a suitable current to operate a sensor and so eliminate the need for a separate power supply.

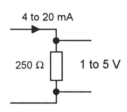

Figure 3.14 *Standard analogue signal*

A potential divider (Figure 3.15) can be used to reduce a voltage from a sensor to the required level; the output voltage level V_{out} is:

$$V_{out} = \frac{R_2}{R_1 + R_2} V_{in}$$

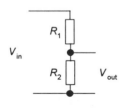

Figure 3.15 *Potential divider*

Amplifiers can be used to increase the voltage level; Figure 3.16 shows the basic form of the circuits that might be used with a 741 operational amplifier with (a) being an inverting amplifier and (b) a non-inverting amplifier. With the inverting amplifier the output V_{out} is:

$$V_{out} = -\frac{R_2}{R_1} V_{in}$$

and with the non-inverting amplifier:

$$V_{out} = \frac{R_1 + R_2}{R_1} V_{in}$$

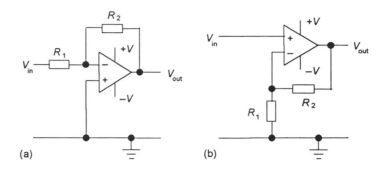

Figure 3.16 *Operational amplifier circuits*

Often a differential amplifier is needed to amplify the difference between two input voltages. Such is the case when a sensor, e.g. a strain gauge, is connected in a Wheatstone bridge and the output is the difference between two voltages or a thermocouple where the voltage difference between the hot and cold junctions is required. Figure 3.17 shows the basic form of an operational amplifier circuit for this purpose. The output voltage V_{out} is:

$$V_{out} = \frac{R_2}{R_1}(V_2 - V_1)$$

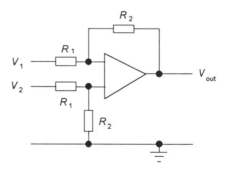

Figure 3.17 *Differential amplifier*

As an illustration of the use of signal conditioning, Figure 3.18 shows the arrangement that might be used for a strain gauge sensor. The sensor is connected in a Wheatstone bridge and the out-of-balance potential difference amplified by a differential amplifier before being fed via an analogue-to-digital converter unit which is part of the analogue input port of the PLC.

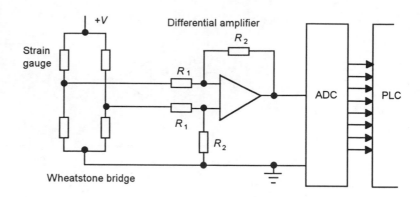

Figure 3.18 *Signal conditioning with a strain gauge sensor*

3.3 Remote input/output connections

When there are many inputs or outputs located considerable distances away from the PLC, while it would be possible to run cables from each such device to the PLC a more economic solution is to use input/output modules in the vicinity of the inputs and outputs and use cables to connect these, over the long distances, to the PLC (Figure 3.19).

In some situations a number of PLCs may be linked together with a master PLC unit sending and receiving input/output data from the other units (Figure 3.20). The distant PLCs do not contain the control program since all the control processing is carried out by the master PLC.

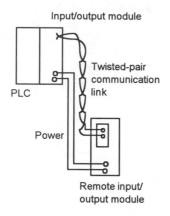

Figure 3.19 *Use of remote input/output module*

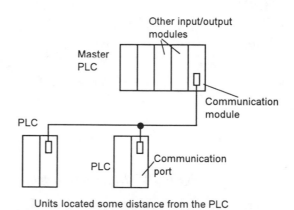

Figure 3.20 *Use of remote input/output PLC systems*

The cables used for communicating data between remote input/output modules and a central PLC, remote PLCs and the master PLC are typically *twisted-pair cabling*, often routed through grounded steel conduit in order to reduce electrical 'noise'. *Coaxial cable* enables higher data rates to be transmitted and does not require the shielding of steel conduit. *Fibre-optic cabling* has the advantage of resistance to noise, small size and flexibility and is now becoming more widely used.

3.3.1 Serial and parallel communications

Serial communication is when data is transmitted one bit at a time. Thus if an 8-bit word is to be transmitted, the eight bits are transmitted one at a time in sequence along a cable. This means that a data word has to be separated into its constituent bits for transmission and then reassembled into the word when received. Serial communication is used for transmitting data over long distances. *Parallel communication* is when all the constituent bits of a word are simultaneously transmitted along parallel cables. This allows data to be transmitted over short distances at high speeds. With a PLC system, serial communication might be used for the connection between a computer, when used as a programming terminal, and a PLC. Parallel communication might be used when connecting laboratory instruments to the system.

3.2.2 Serial standards

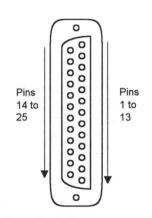

Pins 14 to 25

Pins 1 to 13

Figure 3.21 *D connector*

The most common standard serial communications interface used is the *RS232*. Connections are made via 25-pin D-type connectors (Figure 3.21) with usually, though not always, a male plug on cables and a female socket on the equipment. Not all the pins are used in every application. The minimum requirements are:

Pin 1: Ground connection to the frame of chassis
Pin 2: Serial transmitted data (output data pin)
Pin 3: Serial received data (input data pin)
Pin 7: Signal ground which acts as a common signal return path

A configuration that is widely used with interfaces involving computers is:

Pin 1:   Ground connection to the frame of chassis
Pin 2:   Serial transmitted data (output data pin)
Pin 3:   Serial received data (input data pin)
Pin 4:   Request to send
Pin 5:   Clear to send
Pin 6:   Data set ready
Pin 7:   Signal ground which acts as a common signal return path
Pin 20: Data terminal ready

The signals sent through pins 4, 5, 6 and 20 are used to check that the receiving end is ready to receive a signal, the transmitting end is ready to send and the data is ready to be sent. With RS232, a 1 bit is represented by a voltage between −5 and −25 V, normally −12 V, and a 0 by a voltage between +5 and +25 V, normally +12 V.

Other standards such as the *RS422* and *RS423* are similar to RS232 although they permit higher transmission rates and longer cable distances.

The term *baud rate* is used to describe the transmission rate, it being approximately the number of bits transmitted or received per second. However, not all the bits transmitted can be used for data, some have to be used to indicate the start and stop of a serial piece of data, these often being termed *flags*, and as a check as to whether the data has been corrupted during transmission. Figure 3.22 shows the type of signal that might be sent with RS232. The parity bit is added to check whether corruption has occurred, with even parity a 1 being added to make the number of 1s an even number. To send seven bits of data, eleven bits may be required.

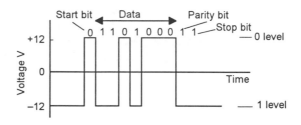

Figure 3.22 *RS232 signal levels*

RS232 is limited over the distances it can be used, noise limiting the transmission of high numbers of bits per second when the length of cable is more than about 15 m. RS422 can be used for longer distances. This uses a balanced method of transmission. Such circuits require two lines for the transmission, the transmitted signal being the voltage difference between the two lines. Noise affecting both lines equally will have no effect on the transmitted signal. Figure 3.23 shows how, for RS232 and RS422, the data rates that can be transmitted without noise becoming too significant depend on the distance. RS422 lines can be used for much greater distances than RS232.

An alternative to RS422 is the *20 mA loop* which was an earlier standard and is still widely used for long distance serial communication, particularly in industrial systems where the communication path is likely to suffer from electrical noise (Figure 3.24). This system consists of a circuit, a loop of wire, containing a current source. The serial data is transmitted by the current being switched on and off, a 0 being transmitted as zero current and a 1 as 20 mA.

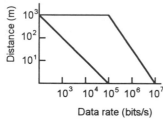

Figure 3.23 *Transmission with RS232 and RS422*

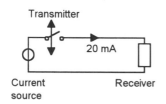

Figure 3.24 *20 mA loop*

3.3.3 Parallel standards

The standard interface most commonly used for parallel communications is *IEEE-488*. This was originally developed by Hewlett Packard to link their computers and instruments and was known as the *Hewlett Packard Instrumentation Bus*. It is now often termed the *General Purpose Instrument Bus*. This bus provides a means of making interconnections so that parallel data communications can take place between listeners, talkers and controllers. Listeners are devices that accept data from the bus, talkers place data, on request, on the bus and controllers manage the flow of data on the bus and provide processing facilities. There is a total of 24 lines, of which eight bi-directional lines are used to carry data and commands between the various devices connected to the bus, five lines are used for control and status signals, three are used for handshaking between devices and eight are ground return lines (Figure 3.25).

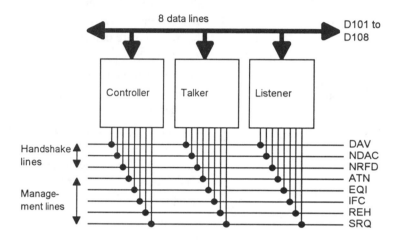

Figure 3.25 *The IEEE-488 bus structure*

Commands from the controller are signalled by taking the Attention Line (ATN) low, otherwise it is high, and thus indicating that the data lines contain data. The commands can be directed to individual devices by placing addresses on the data lines. Each device on the bus has its own address. Device addresses are sent via the data lines as a parallel 7-bit word, the lowest 5-bits providing the device address and the other two bits control information. If both these bits are 0 then the commands are sent to all addresses, if bit 6 is 1 and bit 7 a 0 the addressed device is switched to be a listener, if bit 6 is 0 and bit 7 is 1 then the device is switched to be a talker.

As illustrated above by the function of the ATN line, the management lines each have an individual task in the control of information. The handshake lines are used for controlling the transfer of data. The three lines ensure that the talker will only talk when it is being listened to by listeners. Table 3.1 lists the functions of all the lines and their pin numbers in a 25-way D-type connector.

Table 3.1 *IEEE-488 bus system*

Pin	Signal group	Abbreviation	Signal/function
1	Data	D101	Data line 1
2	Data	D102	Data line 2
3	Data	D103	Data line 3
4	Data	D104	Data line 4
5	Management	EOI	End Or Identify. This is used to either signify the end of a message sequence from a talker device or is used by the controller to ask a device to identify itself.
6	Handshake	DAV	Data valid. When the level is low on this line then the information on the data bus is valid and acceptable.
7	Handshake	NRFD	Not Ready For Data. This line is used by listener devices taking it high to indicate that they are ready to accept data.
8	Handshake	NDAC	Not Data Accepted. This line is used by listeners taking it high to indicate that data is being accepted.
9	Management	IFC	Interface Clear. This is used by the controller to reset all the devices of the system to the start state.
10	Management	SRQ	Service Request. This is used by devices to signal to the controller that they need attention.
11	Management	ATN	Attention. This is used by the controller to signal that it is placing a command on the data lines.
12		SHIELD	Shield.
13	Data	D105	Data line 5.
14	Data	D106	Data line 6.
15	Data	D107	Data line 7.
16	Data	D108	Data line 8.
17	Management	REN	Remote Enable. This enables a device on the bus to indicate that it is to be selected for remote control rather than by its own control panel.
18		GND	Ground/common.
19		GND	Ground/common.
20		GND	Ground/common.
21		GND	Ground/common.
22		GND	Ground/common.
23		GND	Ground/common.
24		GND	Ground/common.

Figure 3.26 shows the handshaking sequence that occurs when data is put on the data lines. Initially DAV is high, indicating that there is no valid data on the data bus, NRFD and NDAC also being low. When a data word is put on the data lines, NRFD is made high to indicate that all listeners are ready to accept data and DAV is made low to indicate that new data is on the data lines. When a device accepts a data word it sets NDAC high to indicate that is has accepted the data and NRFD low to indicate that it is now not ready to accept data. When all the listeners have set NDAC high, then the talker cancels the data valid signal, DAV going high. This then results in NDAC being set low. The entire process can then be repeated for another word being put on the data bus.

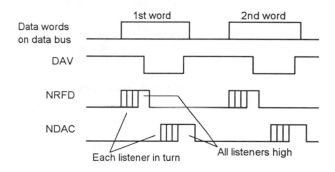

Figure 3.26 *Handshaking sequence*

3.3.4 Protocols

It is necessary to exercise control of the flow of data between two devices so what constitutes the message, and how the communication is to be initiated and terminated, is defined. This is termed the *protocol*.

Thus one device needs to indicate to the other to start or stop sending data. This can be done by using handshaking wires connecting transmitting and receiving devices so that a signal along one such wire can tell the receiver that the transmitter is ready to send (RTS) and along another wire that the transmitter is ready to receive, a clear to send signal (CTS). RTS and CTS lines are provided for in RS232 serial communication links.

An alternative is to use additional characters on the transmitting wires. With the ENQ/ACK protocol, data packets are sent to a receiver with a query character ENQ. When this character is received the end of the data packet has been reached. Once the receiver has processed that data, it can indicate it is ready for another block of data by sending back an acknowledge (ACK) signal. Another form, the XON/XOFF, has the receiving device sending a XOFF signal to the sending device when it wishes the data flow to cease. The transmitter then waits for an XON signal before resuming transmission.

One form of checking for errors in the message that might occur as a result of transmission is the *parity check*. This is an extra bit added to a message to ensure that the number of bits in a piece of data is always odd or always even. For example, 0100100 is even since there is an even number of 1s and 0110100 is odd since there is an odd number of 1s. To make both these odd parity then the extra bit added at the end in the first case is 1 and in the second case 0, i.e. we have 01001001 and 01101000. Thus when a message is sent out with odd bit parity, if the receiver finds that the bits give an even sum, then the message has been corrupted during transmission and the receiver can request that the message be repeated.

The parity bit method can detect if there is an error resulting from a single 0 changing to a 1 or a 1 changing to a 0 but cannot detect two such errors occurring since there is then no change in parity. To check on such occurrences more elaborate checking methods have to be used. One method involves storing data words in an array of rows and columns. Parity can then be checked for each row and each column. The following illustrates this for seven words using even parity.

		Row parity bits
Columns parity bits	00101010	1
↑	10010101	0
	10100000	0
Block	01100011	0
of data	11010101	1
	10010101	1
↓	00111100	0

Another method, termed *cyclic redundancy check codes*, involves splitting the message into blocks. Each block is then treated as a binary number and is divided by a predetermined number. The remainder from this division is then sent as the error checking number on the conclusion of the message and enables a check on the accuracy of the message to be undertaken.

3.4 Networks

The increasing use of automation in industry has led to the need for communications and control on a plant-wide basis with programmable controllers, computers, robots, CNC machines interconnected. The term *local area network (LAN)* is used to describe a communications network designed to link computers and their peripherals within the same building or site. Networks can take three basic forms. With the *star* form (Figure 3.27(a)) the terminals are each directly linked to a central computer, termed the host, or master with the terminals being termed slaves. The host contains the memory, processing and switching equipment to enable the terminals to communicate. Access to the terminals is by the host asking each terminal in turn whether it wants to talk or listen. With the *bus* or *single highway* type of network (Figure 3.27(b)), each of the terminals is linked into a single cable and so each transmitter/receiver has a direct path to each other transmitter/receiver in the network. Methods, i.e. protocols, have to be adopted to ensure that no more than one terminal talks at once, otherwise confusion can occur. A terminal has to be able to detect whether another terminal is talking before it starts to talk. With the *ring* network (Figure 3.27(c)), a continuous cable, in the form of a ring, links all the terminals. Again methods have to be employed to enable communications from different terminals without messages becoming mixed up. The single highway and the ring methods are often termed *peer to peer* in that each terminal has equal status.

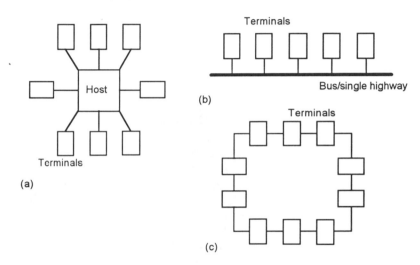

Figure 3.27 *Networks: (a) star, (b) bus/single highway, (c) ring*

Different PLC manufacturers adopt different forms of network systems and methods of communication for use with their PLCs. For example, Mitsubishi uses a network termed MelsecNET, Allen Bradley uses Data Highway Plus, General Electric uses GENET, Texas Instruments uses TIWAY and Siemens has four forms under the general name SINEC. Most employ peer to peer forms, e.g. Allen Bradley. Siemens has two low level forms, SINECLI which is a star, i.e. master-slave, form and SINECL2 which is peer to peer.

Often PLCs figure in an entire hierarchy of communications (Figure 3.28). Thus at the lowest level we have input and output devices such as sensor and motors interfaced through input/output interfaces with the next level. The next level involves controllers such as small PLCs or small computers, linked through a network with the next level of larger PLCs and computers exercising local area control. These in turn may be part of a network involved with a large mainframe company computer controlling all.

There is increasing use made of systems that can both control and monitor industrial processes. This involves control and the gathering of data. The term SCAD, which stands for *supervisory control and data acquisition system*, is widely used for such a system.

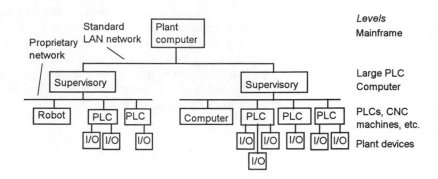

Figure 3.28 *Control hierarchy*

3.4.1 Network standards

Interconnecting several devices can present problems because of compatibility problems, e.g. they may operate at different baud rates or use different protocols. In order to facilitate communications between different devices the International Standards Organisation (ISO) in 1979 devised a model to be used for standardisation for open systems interconnection (OSI); the model is termed the *ISO/OSI model*. A communication link between items of digital equipment is defined in terms of physical, electrical, protocol and user standards, the ISO/OSI model breaking this down into seven layers (Figure 3.29).

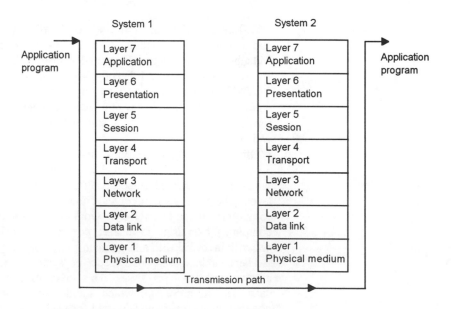

Figure 3.29 *ISO/OSI model*

The function of each layer in the model is:

Layer 1: Physical

This layer is concerned with the coding and physical transmission of information. Its functions include synchronising data transfer and transferring bits of data between systems.

Layer 2: Data link

This layer defines the protocols for sending and receiving information between systems that are directly connected to each other. Its functions include assembling bits from the physical layer into blocks and transferring them, controlling the sequence of data blocks and detecting and correcting errors.

Layer 3: Network

This layer defines the switching that routes data between systems in the network.

Layer 4: Transport

This layer defines the protocols responsible for sending messages from one end of the network to the other. It controls message flow.

Layer 5: Session

This layer provides the function to set up communications between users at separate locations.

Layer 6: Presentation

This layer assures that information is delivered in an understandable form.

Layer 7: Application

This layer has the function of linking the user program into the communication process and is concerned with the meaning of the transmitted information.

Each layer is self contained and only deals with the interfaces of the layer immediately above and below it; it performs its tasks and transfers its results to the layer above or the layer below. It thus enables manufacturers of products to design products operable in a particular layer that will interface with the hardware of other manufacturers.

In 1980, the IEEE (Institute of Electronic and Electrical Engineers) began Project 802. This is a model which adheres to the OSI Physical layer but subdivided the Data link layer into two separate layers: the Media Access Control (MAC) layer and the Logical Link Control (LLC) layer. The MAC layer defines the access method to the transmission medium and consists of a number of standards to control access to the network and ensure that only one user is able to transmit at any one time. One standard is IEEE 802.3 Carrier Sense Multiple Access and Collision Detection (CSMA/CD); stations have to listen for other transmissions before being able to gain control of the network and transmit. Another standard is IEEE 802.4 Token Passing Bus; with this method a special bit pattern, the token, is circulated and when a station wishes to transmit it waits until it receives the token and then attaches it to the end of the data. The LCC layer is responsible for the reliable transmission of data packets across the Physical layer.

General Motors in the United States had a problem in automating their manufacturing activities by 1990 of requiring all their systems to be able to talk to each other. They thus developed a standard communications system for factory automation applications, this being termed the *manufacturing automation protocol* (MAP). The system was for all systems on the shop floor, e.g. robot systems, PLCs, welding systems. Table 3.2 shows the MAP model and its relationship to the ISO model. In order for non-OSI equipment to operate on the MAP system, gateways may be used. These are self-contained units or interface boards that fit in the device so that messages from a non-OSI network/device may be transmitted through the MAP broad band token bus to other systems.

Table 3.2 *MAP*

ISO layer		MAP protocol
7	Application	ISO file transfer, MMFS, FTAM, CASE
6	Presentation	
5	Session	ISO session kernel
4	Transport	ISO transport class 4
3	Network	ISO Internet
2	Data link	IEEE 802.2 class 1; IEEE 802.4 token bus
1	Physical	IEEE 802.4 broad band
	Transmission	10 mbps coaxial cable with RF modulators

Note: MMFS = manufacturing message format standard, FTAM = file transfer, CASE = common applications service; each of these provides a set of commands that will be understood by devices and the software used. For the data link, methods are needed to ensure that only user of the network is able to transmit at any one time and for MAP the method used is token passing. The term broad band is used for a network in which information is modulated onto a radio frequency carrier which is then transmitted through the coaxial cable.

3.5 Processing inputs

A PLC is continuously running through its program and updating it as a result of the input signals. Each such loop is termed a *cycle*. Figure 3.30 illustrates this action. There are two methods that can be used for the input/output processing operations.

1 *Continuous updating*
   This involves the CPU scanning the input channels as they occur in the program instructions. Each input point is examined individually and its effect on the program determined. This interrogation of each input in

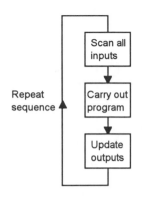

Figure 3.30 *PLC operation*

relation to each program instruction takes time. A number of inputs may have to be scanned before the program has the instruction for some particular operation to be executed and an output to occur. The outputs retain their status, i.e. the term latched is used, until the next updating. The sequence is thus:

> Fetch and decode first program instruction
> Scan the relevant inputs
> Fetch and decode second program instruction
> Scan the relevant inputs, etc. for the remaining program instructions
> Update outputs
> Repeat the entire sequence

2   *Mass input/output copying*
Because, with continuous updating, there is time spent interrogating each input in turn, the time taken to examine several hundred input/output points can become comparatively long. To allow a more rapid execution of a program, a specific area of RAM is used as a buffer store between the control logic and the input/output unit. Each input/output has an address in this memory. At the start of each program cycle the CPU scans all the inputs and copies their status into the input/output addresses in RAM. As the program is executed the stored input data is read, as required, from RAM and the logic operations carried out. The resulting output signals are stored in the reserved input/output section of RAM. At the end of each program cycle all the outputs are transferred from RAM to the appropriate output channels. The outputs retain their status until the next updating. The sequence is thus:

> Scan all the inputs and copy into RAM
> Fetch and decode and execute all program instructions in sequence, copying output instructions to RAM
> Update all outputs
> Repeat the sequence

The time taken to complete a cycle of scanning inputs and updating outputs according to the program instructions, though relatively quick, is not instantaneous and means that the inputs are not watched all the time but samples of their states taken periodically. A typical cycle time is of the order of 10 to 50 ms. This means that the inputs and outputs are updated every 10 to 50 ms and thus there can be a delay of this order in the system reacting. It also means that if a very brief input cycle appears at the wrong moment in the cycle, it could be missed. In general, any input must be present for longer than the cycle time. Special modules are available for use in such circumstances.

Consider a PLC with a cycle time of 40 ms. What is the maximum frequency of digital impulses that can be detected? The maximum frequency will be if one pulse occurs every 40 ms, i.e. a frequency of $1/0.04 = 25$ Hz.

3.6 Input and output addresses

The PLC has to be able to identify each particular input and output. It does this by allocating addresses to each input and output. With a small PLC this is likely to be just a number, prefixed by a letter to indicate whether it is an input or an output. Thus for the Mitsubishi PLC we might have inputs with addresses X400, X401, X402, etc. and outputs with addresses Y430, Y431, Y432, etc., the X indicating an input and the Y an output. Toshiba use a similar system.

With larger PLCs having several racks of input and output channels, the racks are numbered. With the Allen Bradley PLC-5, the rack containing the processor is given the number 0 and the addresses of the other racks are numbered 1, 2, 3, etc. according to how set-up switches are set. Each rack can have a number of modules and each one deals with a number of inputs and/or outputs. Thus addresses can be of the form shown in Figure 3.31. For example, we might have an input with address I:012/03. This would indicate an input, rack 01, module 2 and terminal 03.

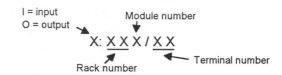

Figure 3.31 *Allen Bradley PLC-5 addressing*

With the Siemens SIMATIC S5, the inputs and outputs are arranged in groups of 8. Each 8 group is termed a byte and each input or output with an 8 is termed a bit. The inputs and outputs thus have their addresses in terms of the byte and bit numbers, effectively giving a module number followed by a terminal number, a full stop (.) separating the two numbers. Figure 3.32 shows the system. Thus I0.1 is an input at bit 1 in byte 0, Q2.0 is an output at bit 0 in byte 2.

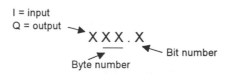

Figure 3.32 *Siemens SISMATIC S5 addressing*

The GEM-80 PLC has its inputs and output addresses in terms of the module number and terminal number within that module. The letter A is used to designate inputs and B outputs. Thus A3.02 is an input at terminal 02 in module 3, B5.12 is an output at terminal 12 in module 5.

With the Sprecher+Schuh SESTEP, inputs are designated by X and outputs by Y and just numbered sequentially, the position of the module concerned not affecting the address. Thus we might have X002 for an input and Y003 for an output.

In addition to using addresses to identify inputs and outputs, PLCs also use their addressing systems to identify internal, software-created devices, such as relays, timers and counters.

Problems

Questions 1 to 9 have four answer options: A, B, C or D. Choose the correct answer from the answer options.

1 Decide whether each of these statements is True (T) or False (F).

A serial communication interface:
(i) Involves data being transmitted and received one bit at a time.
(ii) Is a faster form of transmission than parallel communication.

A (i) T (ii) T
B (i) T (ii) F
C (i) F (ii) T
D (i) F (ii) F

2 Decide whether each of these statements is True (T) or False (F).

The RS232 communications interface:
(i) Is a serial interface.
(ii) Is typically used for distances up to about 15 m.

A (i) T (ii) T
B (i) T (ii) F
C (i) F (ii) T
D (i) F (ii) F

Questions 3 and 4 refer to the following which shows the bits on an RS232 data line being used to transmit the data 1100001:

```
0110000111
X       YZ
```

3 Decide whether each of these statements is True (T) or False (F).

The extra bits X and Z at the beginning and the end are:
(i) To check whether the message is corrupted during transmission.
(ii) To indicate where the data starts and stops.

A (i) T (ii) T
B (i) T (ii) F
C (i) F (ii) T
D (i) F (ii) F

4 Decide whether each of these statements is True (T) or False (F).

Bit Y is:
(i) The parity bit showing odd parity.
(ii) The parity bit showing even parity.

A (i) T (ii) T
B (i) T (ii) F
C (i) F (ii) T
D (i) F (ii) F

5 Decide whether each of these statements is True (T) or False (F).

The parallel data communication interface:
(i) Enables data to be transmitted over short distances at high speeds.
(ii) A common standard is IEEE-488.

A (i) T (ii) T
B (i) T (ii) F
C (i) F (ii) T
D (i) F (ii) F

6 Decide whether each of these statements is True (T) or False (F).

For communications over distances of the order of 100 to 300 m with a high transmission rate:
(i) The RS232 interface can be used.
(ii) The 20 mA current loop can be used.

A (i) T (ii) T
B (i) T (ii) F
C (i) F (ii) T
D (i) F (ii) F

7 Decide whether each of these statements is True (T) or False (F).

With input/output processing, mass input/output copying:
(i) Scans all the inputs and copies their states into RAM.
(ii) Is a faster process than continuous updating.

A (i) T (ii) T
B (i) T (ii) F
C (i) F (ii) T
D (i) F (ii) F

8 The cycle time of a PLC is the time it takes to:

A Read an input signal.
B Read all the input signals.
C Check all the input signals against the program.
D Read all the inputs, run the program and update all outputs.

9 Decide whether each of these statements is True (T) or False (F).

A PLC with a long cycle time is suitable for:
(i) Short duration inputs.
(ii) High frequency inputs.

A (i) T   (ii) T
B (i) T   (ii) F
C (i) F   (ii) T
D (i) F   (ii) F

10 Specify (a) the odd parity bit, (b) the even parity bit, to be used when the data 1010100 is transmitted.

11 Explain the purpose of using a parity bit.

12 Explain the continuous updating and the mass input/output copying methods of processing inputs/outputs.

4 Programming

Programs for microprocessor-based systems have to be loaded into them in *machine code*, this being a sequence of binary code numbers to represent the program instructions. However, *assembly language* based on the use of mnemonics can be used, e.g. LD is used to indicate the operation required to load the data that follows the LD, and a computer program called an assembler is used to translate the mnemonics into machine code. Programming can be made even easier by the use of the so-called *high level languages*, e.g. C, BASIC, PASCAL, FORTRAN, COBOL. These use pre-packaged functions, represented by simple words or symbols descriptive of the function concerned. For example, with C language the symbol & is used for the logic AND operation. However, the use of these methods to write programs requires some skill in programming and PLCs are intended to be used by engineers without any great knowledge of programming. As a consequence, ladder programming was developed. This is a means of writing programs which can then be converted into machine code by some software for use by the PLC microprocessor.

This chapter is an introduction to the programming of a PLC using ladder diagrams, with just a brief mention of other techniques. It is concerned with the basic techniques involved in developing such programs to represent basic switching operations, involving the logic functions of AND, OR, Exclusive OR, NAND and NOR. The chapters that follow continue with further programming.

4.1 Ladder diagrams

As an introduction to ladder diagrams, consider the simple wiring diagram for an electrical circuit in Figure 4.1(a). The diagram shows the circuit for switching on or off an electric motor. We can redraw this diagram in a different way, using two vertical lines to represent the input power rails and stringing the rest of the circuit between them. Figure 4.1(b) shows the result. Both circuits have the switch in series with the motor and supplied with electrical power when the switch is closed. The circuit shown in Figure 4.1(b) is termed a *ladder diagram*.

With such a diagram the power supply for the circuits is always shown as two vertical lines with the rest of the circuit as horizontal lines. The power lines, or rails as they are often termed, are like the vertical sides of a ladder with the horizontal circuit lines like the rungs of the ladder. The horizontal rungs show only the control portion of the circuit, in the case of Figure 4.1 it is just the switch in series with the motor. Circuit diagrams often show the relative physical location of the circuit components and how they are actually wired. With ladder diagrams no attempt is made to show the actual physical locations and the emphasis is on clearly showing how the control is exercised.

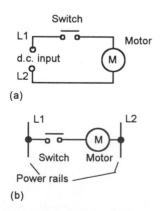

Figure 4.1 *Ways of drawing the same electrical circuit*

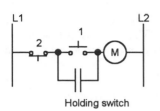

Figure 4.2 *Stop-start circuit*

Figure 4.2 shows an example of a ladder diagram for a circuit that is used to start and stop a motor using push buttons. In the normal state, push button 1 is open and push button 2 closed. When button 1 is pressed, the motor circuit is completed and the motor starts. Also, the holding contacts wired in parallel with the motor close and remain closed as long as the motor is running. Thus when the push button 1 is released, the holding contacts maintain the circuit and hence the power to the motor. To stop the motor, button 2 is pressed. This disconnects the power to the motor and the holding contacts open. Thus when push button 2 is released, there is still no power to the motor. Thus we have a motor which is started by pressing button 1 and stopped by pressing button 2.

4.1.1 PLC ladder programming

A very commonly used method of programming PLCs is based on the use of *ladder diagrams*. Writing a program is then equivalent to drawing a switching circuit. The ladder diagram consists of two vertical lines representing the power rails. Circuits are connected as horizontal lines, i.e. the rungs of the ladder, between these two verticals.

In drawing a ladder diagram, certain conventions are adopted:

1 The vertical lines of the diagram represent the power rails between which circuits are connected.

2 Each rung on the ladder defines one operation in the control process.

3 A ladder diagram is read from left to right and from top to bottom, Figure 4.3 showing the scanning motion employed by the PLC. The top rung is read from left to right. Then the second rung down is read from left to right and so on. When the PLC is in its run mode, it goes through the entire ladder program to the end, the end rung of the program being clearly denoted, and then promptly resumes at the start (see section 3.4). This procedure of going through all the rungs of the program is termed a *cycle*.

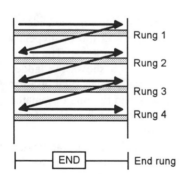

Figure 4.3 *Scanning the ladder program*

4 Each rung must start with an input or inputs and must end with at least one output. The term input is used for a control action, such as closing the contacts of a switch, used as an input to the PLC. The term output is used for a device connected to the output of a PLC, e.g. a motor.

5 Electrical devices are shown in their normal condition. Thus a switch which is normally open until some object closes it, is shown as open on the ladder diagram. A switch that is normally closed is shown closed.

6 A particular device can appear in more than one rung of a ladder. For example, we might have a relay which switches on one or more devices. The same letters and/or numbers are used to label the device in each situation.

7   The inputs and outputs are all identified by their addresses, the notation used depending on the PLC manufacturer. This is the address of the input or output in the memory of the PLC. The Mitsubishi F series of PLCs precedes input elements by an X and output elements by a Y and uses the following numbers:

Inputs   X400–407, 410–413, 500–507, 510–513
(24 possible inputs)

Outputs   Y430–437, 530–537
(16 possible outputs)

Toshiba also uses an X and Y with inputs such as X000 and X001, outputs Y000 and Y001. Siemens precedes input numbers by I and outputs by Q, e.g. I0.1 and Q2.0. Sprecher+Schuh precedes input numbers by X and output numbers by Y, e.g. X001 and Y001. Allen Bradley uses I and O, e.g. I:21/01 and O:22/01. See section 3.5 for a discussion of such addresses.

Figure 4.4 shows standard symbols that are used for input and output devices. Note that inputs are represented by just two symbols representing normally open or normally closed contacts. This applies whatever the form of the device connected to the input. The action of the input is equivalent to opening or closing a switch. Outputs are represented by just one symbol, regardless of the device connected to the output. Further symbols will be introduced in later chapters.

To illustrate the drawing of the rung of a ladder diagram, consider a situation where the energising of an output device, e.g. a motor, depends on a normally open start switch being activated by being closed. The input is thus the switch and the output the motor. Figure 4.5 shows the ladder diagram. Starting with the input, we have the normally open symbol ‖ for the input contacts. There are no other input devices and the line terminates with the output, denoted by the symbol O. When the switch is closed, i.e. there is an input, the output of the motor is activated.

In drawing ladder diagrams the addresses of each element are appended to its symbol. Thus Figure 4.6 shows how the ladder diagram of Figure 4.5 would appear using (a) Mitsubishi, (b) Siemens, (c) Allen Bradley, (d) Telemecanique notations for the addresses. Thus Figure 4.6(a) indicates that this rung of the ladder program has an input from address X400 and an output to address Y430. When wiring up the inputs and outputs to the PLC, the relevant ones must be connected to the input and output terminals with these addresses.

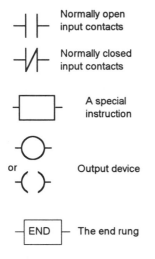

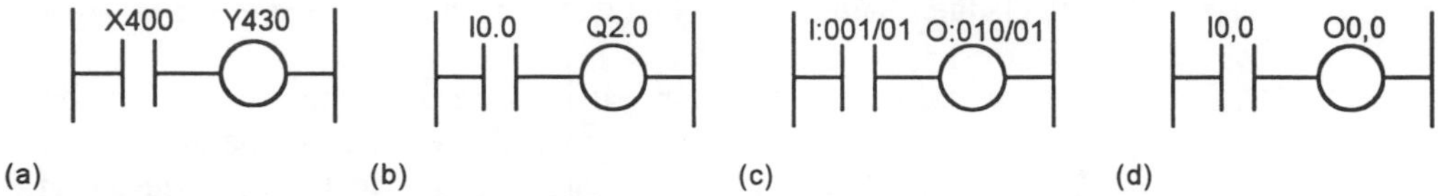

Normally open input contacts

Normally closed input contacts

A special instruction

or   Output device

END   The end rung

Figure 4.4   *Basic symbols*

Input   Output

Figure 4.5   *A ladder rung*

X400   Y430

I0.0   Q2.0

I:001/01   O:010/01

I0,0   O0,0

(a)   (b)   (c)   (d)

Figure 4.6 *Notation: (a) Mitsubishi, (b) Siemens, (c) Allen Bradley, (d) Telemecanique*

4.2 Logic functions

There are many control situations requiring actions to be initiated when a certain combination of conditions is realised. Thus, for an automatic drilling machine (as illustrated in Figure 1.1(a)), there might be the condition that the drill motor is to be activated when the limit switches are activated that indicate the presence of the workpiece and the drill position as being at the surface of the workpiece. Such a situation involves the AND logic function, condition A and condition B having both to be realised for an output to occur. This section is a consideration of such logic functions.

4.2.1 AND

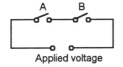

Figure 4.7 *AND circuit*

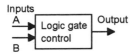

Figure 4.8 *AND logic gate*

Figure 4.7 shows a situation where an output is not energised unless two, normally open, switches are both closed. Switch A and switch B have both to be closed, which thus gives an AND logic situation. We can think of this as representing a control system with two inputs A and B (Figure 4.8). Only when A and B are both on is there an output. Thus if we use 1 to indicate an on signal and 0 to represent an off signal, then for there to be a 1 output we must have A and B both 1. Such an operation is said to be controlled by a *logic gate* and the relationship between the inputs to a logic gate and the outputs is tabulated in a form known as a *truth table*. Thus for the AND gate we have:

Inputs		Output
A	B	
0	0	0
0	1	0
1	0	0
1	1	1

Figure 4.9 shows an AND gate system on a ladder diagram. The ladder diagram starts with ‖, a normally open set of contacts labelled input A, to represent switch A and in series with it ‖, another normally open set of contacts labelled input B, to represent switch B. The line then terminates with O to represent the output. For there to be an output, both input A and input B have to occur, i.e. input A and input B contacts have to be closed. Figure 4.10 shows Figure 4.9 in terms of (a) Mitsubishi, (b) Siemens and (c) Toshiba address notations.

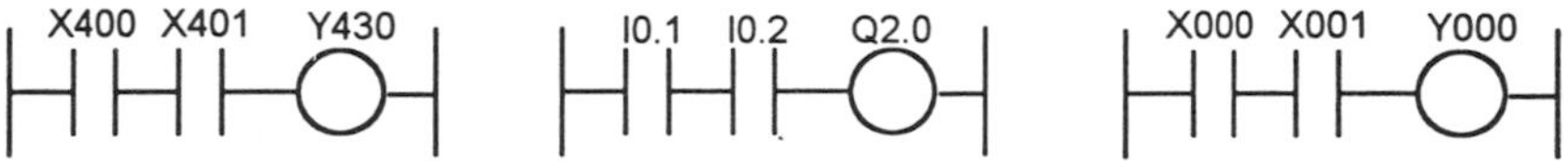

Figure 4.9 *AND gate*

Figure 4.10 *AND gate: (a) Mitsubishi, (b) Siemens, (c) Toshiba notations*

An example of an AND gate is an interlock control system for a machine tool so that it can only be operated when the safety guard is in position and the power switched on.

4.2.2 OR

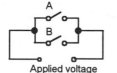

Figure 4.11 *OR circuit*

Figure 4.11 shows a situation where an output is energised when switch A or B, both normally open, is closed. This describes an OR logic gate in that input A <u>or</u> input B must be on for there to be an output. The truth table is:

Inputs		Output
A	B	
0	0	0
0	1	1
1	0	1
1	1	1

Figure 4.12(a) shows an OR logic gate system on a ladder diagram, Figure 4.12(b) showing an equivalent alternative way of drawing the same diagram. The ladder diagram starts with ||, normally open contacts labelled input A, to represent switch A and in parallel with it ||, normally open contacts labelled input B, to represent switch B. Either input A <u>or</u> input B have to be closed for the output to be energised. The line then terminates with O to represent the output. Figure 4.13 shows how such a gate could appear with Mitsubishi, Siemens and Sprecher+Schuh notations.

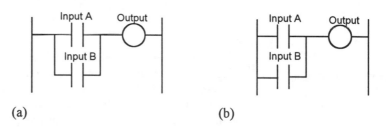

(a)   (b)

Figure 4.12 *OR gate*

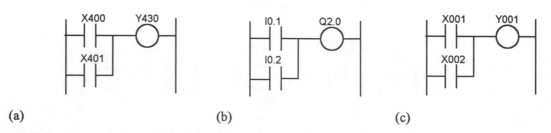

(a)   (b)   (c)

Figure 4.13 *OR gate: (a) Mitsubishi, (b) Siemens, (c) Sprecher+Schuh notations*

An example of an OR gate control system is a conveyor belt transporting bottled products to packaging where a deflector plate is activated to deflect bottles into a reject bin if either the weight is not within certain tolerances or there is no cap on the bottle.

4.2.3 NOT

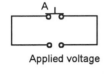

Figure 4.14 *NOT circuit*

Figure 4.14 shows an electrical circuit controlled by a switch that is normally closed. When there is an input to the switch, it opens and there is then no current in the circuit. This illustrates a NOT gate in that there is an output when there is no input and no output when there is an input. The gate is sometimes referred to an *inverter*. The truth table is:

Input A	Output
0	1
1	0

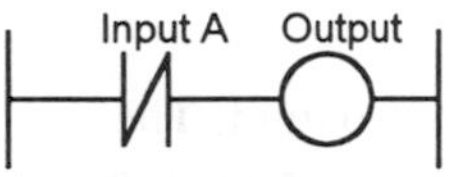

Figure 4.15 *NOT gate*

Figure 4.15 shows a NOT gate system on a ladder diagram. The input A contacts are shown as being normally closed. This is in series with the output O. With no input to input A, the contacts are closed and so there is an output. When there is an input to input A, it opens and there is then no output. Figure 4.16 shows Figure 4.15 in (a) Mitsubishi, (b) Siemens and (c) Telemecanique address notations.

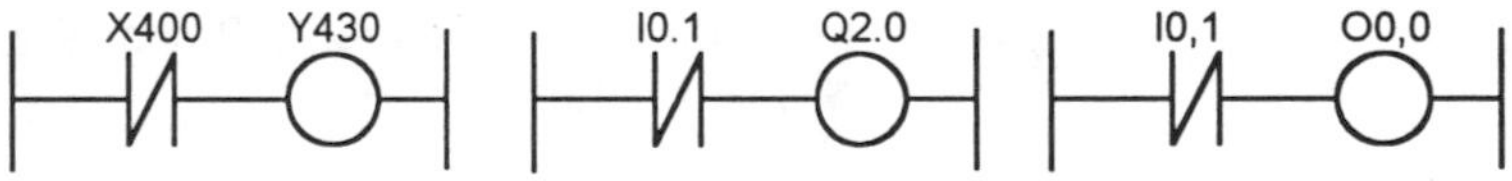

(a)   X400   Y430   (b)   I0.1   Q2.0   (c)   I0,1   O0,0

Figure 4.16 *NOT gate: (a) Mitsubishi, (b) Siemens, (c) Telemecanique notations*

An example of a NOT gate control system is a light that comes on when it becomes dark, i.e. when there is no light input to the light sensor there is an output.

4.2.4 NAND

Suppose we follow an AND gate with a NOT gate (Figure 4.17(a)). The consequence of having the NOT gate is to invert all the outputs from the AND gate. An alternative, which gives exactly the same results, is to put a NOT gate on each input and then follow that with OR (Figure 4.17(b)). The same truth table occurs, namely:

Inputs		Output
A	B	
0	0	1
0	1	1
1	0	1
1	1	0

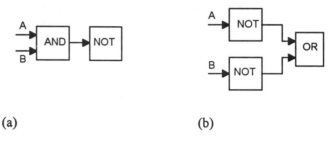

(a)   (b)

Figure 4.17 *NAND gate*

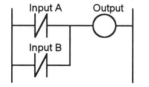

Figure 4.18 *A NAND gate*

Both the inputs A and B have to be 0 for there to be a 1 output. There is an output when input A and input B are not 1. The combination of these gates is termed a NAND gate.

Figure 4.18 shows a ladder diagram which gives a NAND gate. When the inputs to input A and input B are both 0 then the output is 1. When the inputs to input A and input B are both 1, or one is 0 and the other 1, then the output is 0. Figure 4.19 shows Figure 4.18 in (a) Mitsubishi and (b) Siemens notations.

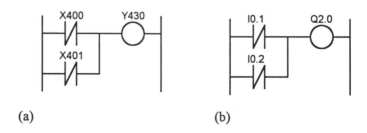

(a)   (b)

Figure 4.19 *NAND gate: (a) Mitsubishi, (b) Siemens notations*

An example of a NAND gate control system is a warning light that comes on if, with a machine tool, the safety guard switch has not been activated and the limit switch signalling the presence of the workpiece has not been activated.

4.2.5 NOR

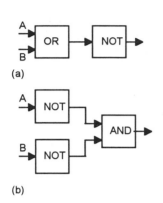

(a)

(b)

Figure 4.20 *NOR gate*

Suppose we follow an OR gate by a NOT gate (Figure 4.20(a)). The consequence of having the NOT gate is to invert the outputs of the OR gate. An alternative, which gives exactly the same results, is to put a NOT gate on each input and then an AND gate for the resulting inverted inputs (Figure 4.20(b)). The following is the resulting truth table:

Inputs		Output
A	B	
0	0	1
0	1	0
1	0	0
1	1	0

The combination of OR and NOT gates is termed a NOR gate. There is an output when neither input A or input B is 1.

Figure 4.21 shows a ladder diagram of a NOR system. When input A and input B are both not activated, there is a 1 output. When either X400 or X401 are 1 there is a 0 output. Figure 4.22 shows the NOR gate system in (a) Mitsubishi, (b) Siemens notations.

Figure 4.21 *NOR gate*

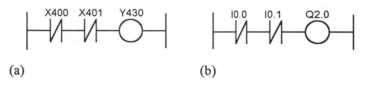

(a)   (b)

Figure 4.22 *NOR gate: (a) Mitsubishi, (b) Siemens notations*

4.2.6 Exclusive OR (XOR)

The OR gate gives an output when either or both of the inputs are 1. Sometimes there is, however, a need for a gate that gives an output when either of the inputs is 1 but not when both are 1, i.e. has the truth table:

Inputs		Outputs
A	B	
0	0	0
0	1	1
1	0	1
1	1	0

Such a gate is called an *Exclusive* OR or XOR gate. One way of obtaining such a gate is by using NOT, AND and OR gates as shown in Figure 4.23.

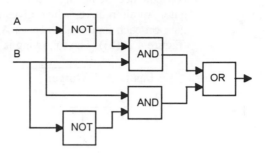

Figure 4.23 *XOR gate*

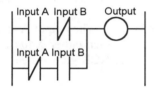

Figure 4.24 *XOR gate*

Figure 4.24 shows a ladder diagram for an XOR gate system. When input A and input B are not activated then there is 0 output. When just input A is activated, then the upper branch results in the output being 1. When just input B is activated, then the lower branch results in the output being 1. When both input A and input B are activated, there is no output. In this example of a logic gate, input A and input B have two sets of contacts in the circuits, one set being normally open and the other normally closed. With PLC programming, each input may have as many sets of contacts as necessary. Figure 4.25 shows Figure 4.24 using (a) Mitsubishi, (b) Siemens, (c) Toshiba, (d) Telemecanique and (e) Allen Bradley addresses.

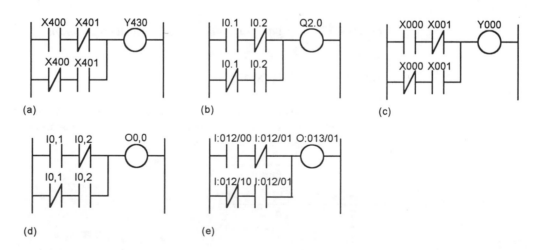

Figure 4.25 *XOR gate: (a) Mitsubishi, (b) Siemens, (c) Toshiba, (d) Telemecanique and (e) Allen Bradley*

4.3 Latching   There are often situations where it is necessary to hold an output energised, even when the input ceases. A simple example of such a situation is a motor which is started by pressing a push button switch. Though the switch contacts do not remain closed, the motor is required to continue running until a stop push button switch is pressed. The term *latch circuit* is used for the circuit used to carry out such an operation. It is a self-maintaining circuit in that, after being energised, it maintains that state until another input is received.

An example of a latch circuit is shown in Figure 4.26, (b) showing the circuit in the Mitsubishi form of addresses. When the input A contacts close, there is an output. However, when there is an output, another set of contacts associated with the output closes. These contacts form an OR logic gate system with the input contacts. Thus, even if the input A opens, the circuit will still maintain the output energised. The only way to release the output is by operating the normally closed contact B.

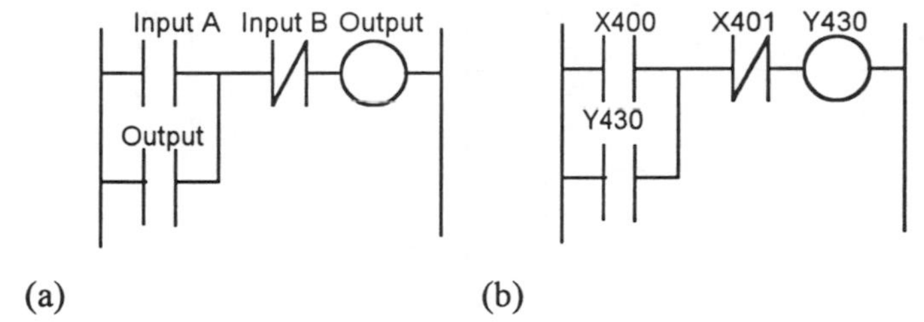

Figure 4.26 *Latched circuit*

As an illustration of the application of a latching circuit, consider a motor controlled by stop and start push button switches and for which one signal light must be illuminated when the power is applied to the motor and another when it is not applied. Figure 4.27 shows the ladder diagram in Mitsubishi notation.

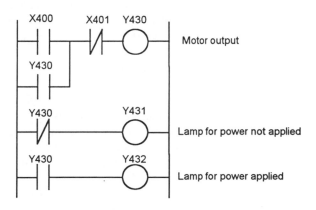

Figure 4.27 *Motor on–off, with signal lamps, ladder diagram*

When X400 is momentarily closed, Y430 is energised and its contacts close. This results in latching and also the switching off of Y431 and the switching on of Y432. To switch the motor off, X401 is pressed and opens. Y430 contacts open in the top rung and third rung, but close in the second rung. Thus Y431 comes on and Y432 off.

4.4 Multiple outputs

With ladder diagrams, there can be more than one output connected to a contact. Figure 4.28 illustrates this with the same ladder program in Mitsubishi and Siemens notations. Outputs Y430, Y431 and Y432 are switched on as the contacts in the sequence given by the contacts X400, X401 and X402 are being closed. Until X400 is closed, none of the other outputs can be switched on. When X400 is closed, Y430 is switched on. Then, when X01 is closed, Y431 is switched on. Finally, when X402 is closed, Y432 is switched on.

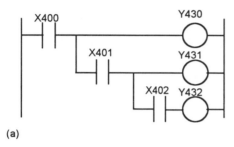

(a)   (b)

Figure 4.28 *Multiple outputs*

4.5 Entering ladder programs

Each horizontal rung on the ladder represents an instruction in the program to be used by the PLC. The entire ladder gives the complete program. There are several methods that can be used for keying in the program into a programming terminal. Whatever method is used to enter the program into a programming terminal or computer, the output to the memory of the PLC has to be in a form that can be handled by the microprocessor. This is termed *machine language* and is just binary code, e.g. 0010100001110001.

4.5.1 Ladder symbols

One method of entering the program into the programming terminal involves using a keypad having keys with symbols depicting the various elements of the ladder diagram and keying them in so that the ladder diagram appears on the screen of the programming terminal. For example, to enter a pair of contacts the key marked

```
┤ ├
```

might be used, followed by its address being keyed in. To enter an output the key marked

might be used, followed by its address. To indicate the start of a junction

might be pressed; to indicate the end of a junction path

To indicate horizontal circuit links, the following key might be used:

The terminal then translates the program drawn on the screen into machine language.

Computers can be used to draw up a ladder program. These involve loading the computer with the relevant software, e.g. MEDOC for Mitsubishi PLCs, and then selecting items from menus on the screen. Thus, from the main menu that appears initially on the screen, Edit is selected. The menu then changes. Ladder can then be selected from this new menu and results in the screen showing a blank ladder diagram consisting of just two parallel rails. At the bottom of the screen a series of ladder symbols appear

```
   1      2       3    4      5   6     7      8     9
  -| |-  -|/|-   -| |-  -|/|-  |   --   -( )-  -[ ]-  P.I
```

and are selected, after pressing F2 to enter the ladder diagrams working area and F7 to select the write mode, by entering the number which appears above the symbol. A window will appear in which the address of the item can be entered and then, on pressing enter, the symbol with its address will appear on the ladder. In this way the entire ladder program can be built up on screen. Text can be inserted, e.g. at the beginning of the program to describe its purpose, by pressing F2 to enter the working area and then F5. This opens a window into which text can be typed.

4.6 Instruction lists   Another programming method, which can be considered to be the entering of a ladder program using text, is *instruction lists*. For this, mnemonic codes are used, each code corresponding to a ladder element. The codes used differ to some extent from manufacturer to manufacturer, though a standard IEC 1131-3 has been proposed. Table 4.1 shows some of the codes used by manufacturers, and the proposed standard, for instructions used in this chapter (see later chapters for codes for other functions).

Whenever a rung is started, it must use a start a rung code. This might be LD, or perhaps A or L or STR, to indicate the rung is starting with open contacts, or LDI, or perhaps LDN or LD NOT or AN or LN or STR NOT, to indicate it is starting with closed contacts. All rungs must end with an output. This might be OUT or =.

The following shows how individual rungs on a ladder are entered using the Mitsubishi mnemonics for the AND gate, shown in Figure 4.29. Step 0 is the start of the rung with LD because it is starting with open contacts. Since the address of the input is X400, the instruction is LD X400. This is followed by another open contacts input and so step 1 involves the instruction AND with the address of the element, thus the instruction is AND X401. The rung terminates with an output and so the instruction OUT is used with the address of the output, i.e. OUT Y430. The single rung of a ladder would thus be entered as:

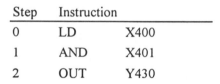

Figure 4.29 *AND gate*

Step	Instruction	
0	LD	X400
1	AND	X401
2	OUT	Y430

Table 4.1 *Instruction code mnemonics*

IEC 1131-3	Mitsubishi	OMRON	Siemens	Telemec- anique	Spreher+ Schuh	
LD	LD	LD	A	L	STR	Start a rung with an open contact
LDN	LDI	LD NOT	AN	LN	STR NOT	Start a rung with a closed contact
AND	AND	AND	A	A	AND	A series element with an open contact
ANDN	ANI	AND NOT	AN	AN	AND NOT	A series element with a closed contact
O	OR	OR	O.	O	OR	A parallel element with an open contact
ORN	ORI	OR NOT	ON	ON	OR NOT	A parallel element with a closed contact
ST	OUT	OUT	=	=	OUT	An output

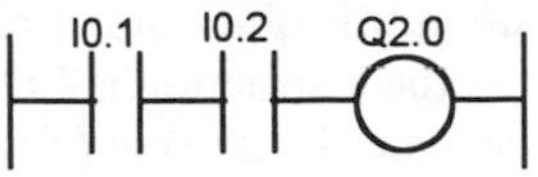

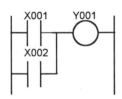

Figure 4.30 *AND gate*

For the same rung with Siemens notation (Figure 4.30) we have:

Step	Instruction	
0	A	I0.1
1	A	I0.2
2	=	Q2.0

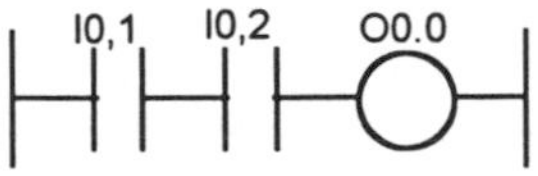

Figure 4.31 *AND gate*

For the same rung with the Telemecanique notation (Figure 4.31) we have:

Step	Instruction	
0	L	I0,1
1	A	I0,2
2	=	O0,0

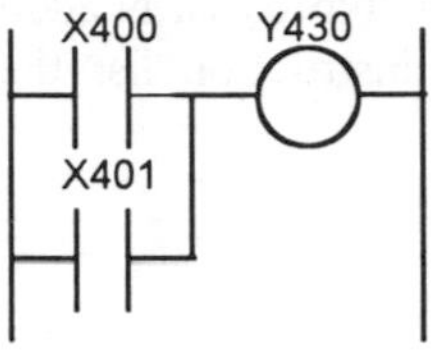

Figure 4.32 *OR gate*

Consider another example, an OR gate. Figure 4.32 shows the gate with Mitsubishi notation. The instructions for the rung start with an open contact is LD X400. The next item is the parallel OR set of contacts X401. Thus the next instruction is OR X401. The last step is the output, hence OUT Y430. The instruction list would thus be:

Step	Instruction	
0	LD	X400
1	OR	X401
2	OUT	Y430

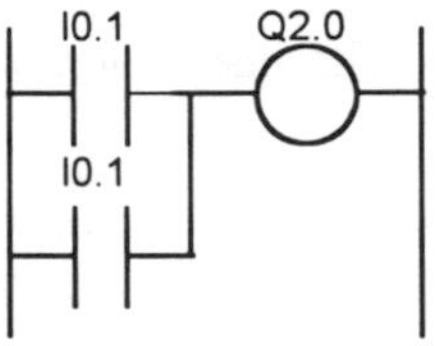

Figure 4.33 *OR gate*

Figure 4.33 shows the Siemens version of the OR gate. The following is the Siemens instruction list:

Step	Instruction	
0	A	I0.1
1	O.	I0.2
2	=	Q2.0

Figure 4.34 *OR gate*

Figure 4.34 shows the Sprecher+Schuh version of the OR gate. The following is thus the Sprecher+Schuh instruction list:

Step	Instruction	
0	STR	X001
1	OR	X002
2	OUT	Y001

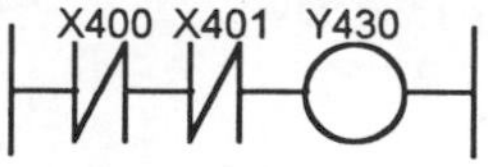

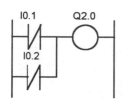

Figure 4.35 *NOR gate*

Figure 4.35 shows the ladder system for a NOR gate in Mitsubishi notation. The rung starts with normally closed contacts and so the instruction is LDI. I when added to Mitsubishi instruction is used to indicate the inverse of the instruction. The next step is a series normally closed contact and so ANI, again the I being used to make an AND instruction the inverse. I is also the instruction for a NOT gate. The instructions for the NOR gate rung of the ladder would thus be entered as:

Step	Instruction	
0	LDI	X400
1	ANI	X401
2	OUT	Y430

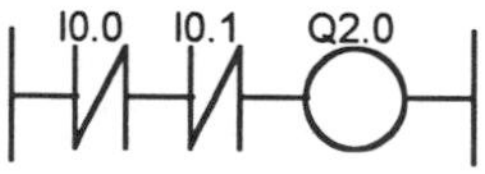

Figure 4.36 *NOR gate*

Figure 4.36 shows the NOR gate with Siemens notation. Note that N added to an instruction is used to make the inverse. The instruction list then becomes:

Step	Instruction	
0	LN	I0.0
1	AN	I0.1
2	=	Q2.0

Figure 4.37 *NOR gate*

Figure 4.37 shows the NOR gate with Sprecher+Schuh notation. Note that NOT is added to an instruction to make the inverse. The instruction list then becomes:

Step	Instruction	
0	STR	X001
1	STR NOT	X002
2	OUT	Y001

Figure 4.38 *NAND gate*

Consider the rung shown in Figure 4.38 in Mitsubishi notation, a NAND gate. It starts with the normally closed contacts X400 and so starts with the instruction LDI X400. The next instruction is for a parallel set of normally closed contacts, thus the instruction is ORI X401. The last step is the output, hence OUT Y430. The instruction list is thus:

Step	Instruction	
0	LDI	X400
1	ORI	X401
2	OUT	Y430

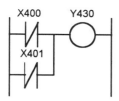

Figure 4.39 *NAND gate*

Figure 4.39 shows the NAND gate in Siemens notation. The instruction list is then:

Step	Instruction	
0	A	I0.1
1	ON	I0.2
2	=	Q2.0

4.6.1 Branch codes

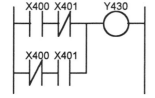

Figure 4.40 *XOR gate*

The Exclusive OR (XOR) gate shown in Figure 4.40 in Mitsubishi notation has two parallel arms with an AND situation in each arm. In such a situation Mitsubishi uses an ORB instruction to indicate 'OR together parallel branches'. The first instruction is for a normally open pair of contacts X400. The next instruction is for a series set of normally closed contacts X401, hence ANI X401. After reading the first two instructions, the third instruction starts a new line. It is recognised as a new line because it starts with LDI, all new lines starting with LD or LDI. But the first line has not been ended by an output. The PLC thus recognises that a parallel line is involved for the second line and reads together the listed elements until the ORB instruction is reached. The mnemonic ORB (OR branches/blocks together) indicates to the PLC that it should OR the results of steps 0 and 1 with that of the new branch with steps 2 and 3. The list concludes with the output OUT Y430. The instruction list would thus be entered as:

Step	Instruction	
0	LD	X400
1	ANI	X401
2	LDI	X400
3	AND	X401
4	ORB	
5	OUT	Y430

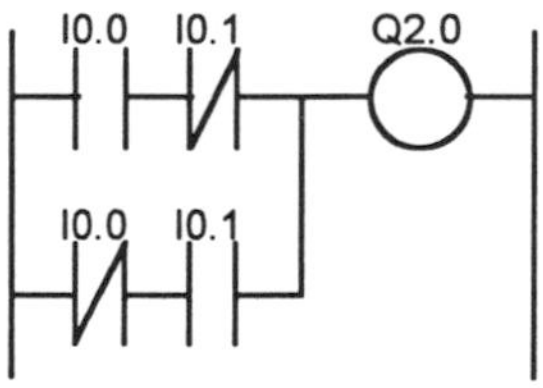

Figure 4.41 *XOR gate*

Figure 4.41 shows the Siemens version of Figure 4.40. Brackets are used to indicate that certain instructions are to be carried out as a block. They are used in the same way as brackets in any mathematical equation. For example, (2 + 3)/4 means that the 2 and 3 must be added before dividing by 4. Thus with the Siemens instruction list we have in step 0 the instruction A(. The brackets close in step 3. This means that the A in step 0 is applied only after the instructions in steps 1 and 2 have been applied.

Step	Instruction		
0	A(		Applies to the first set of brackets
1	A	I0.0	Steps 1 and 2 are in the first set of brackets
2	AN	I0.1	
3	)		

4	O(		Applies to the second set of brackets
5	AN	I0.0	Steps 5 and 6 are in the second set of brackets
6	A	I0.1	
7	)		
8	=	Q2.0	

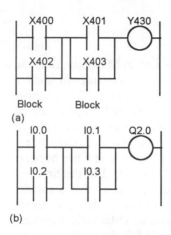

Block   Block

(a)

(b)

Figure 4.42 *(a) Mitsubishi, (b) Siemens*

Figure 4.42(a) shows a circuit, in Mitsubishi notation, which can be considered as two AND blocks. The instruction for this is ANB. The instruction list is thus:

Step	Instruction	
0	LD	X400
1	OR	X402
2	LD	X401
3	OR	X403
4	ANB	
5	OUT	Y430

Figure 4.42(b) shows the same circuit in Siemens notation. Such a program is written as an instruction list using brackets. The A instruction in step 0 applies to the result of steps 1 and 2. The A instruction in step 4 applies to the result of steps 5 and 6. The program instruction list is:

Step	Instruction	
0	A(	
1	A	I0.0
2	O.	I0.1
3	)	
4	A(	
5	A	I0.2
6	O.	I0.3
7	)	
8	=	Q2.0

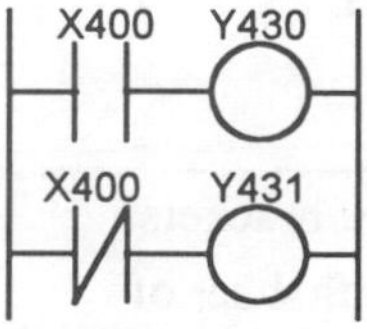

4.6.2 More than one rung

Figure 4.43 shows a ladder, in Mitsubishi notation, with two rungs. In writing the instruction list we just write the instructions for each line in turn. The instruction LD or LDI indicates to the PLC that a new rung is starting. The instruction list is thus:

Figure 4.43 *Toggle circuit*

Step	Instruction	
0	LD	X400
1	OUT	Y430
2	LDI	X400
3	OUT	Y431

The system is one where when X400 is not activated, there is an output from Y431 but not Y430. When X400 is activated, there is then an output from Y430 but not Y431.

Figure 4.44 shows the same program in Siemens notation. The = instruction indicates the end of a line. The A or AN instruction does not necessarily indicate the beginning of a rung since the same instruction is used for AND and AND NOT. The instruction list is then:

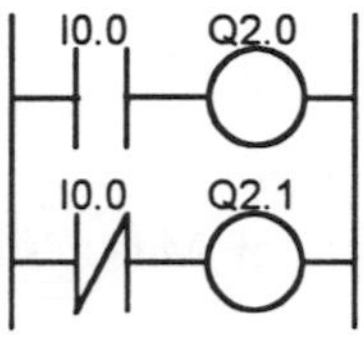

Figure 4.44 *Toggle circuit*

Step	Instruction	
0	A	I0.0
1	=	Q2.0
2	AN	I0.0
3	=	Q2.1

4.7 Boolean algebra

Instruction lists and ladder programs can be derived from Boolean expressions since we are concerned with a mathematical system of logic. In Boolean algebra there are just two digits 0 and 1. When we have an AND operation for inputs A and B then we can write:

$$A \cdot B = Q$$

where Q is the output. Thus Q is equal to 1 only when A = 1 and B = 1. The OR operation for inputs A and B is written as:

$$A + B = Q$$

Thus Q is equal to 1 only when A = 1 or B = 1. The NOT operation for an input A is written as:

$$\overline{A} = Q$$

Thus when A is not 1 there is an output.

As an illustration of how we can relate Boolean expressions with ladder diagrams, consider the expression:

$$A + B \cdot C = Q$$

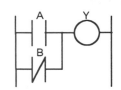

Figure 4.45 *Ladder diagram*

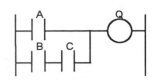

Figure 4.46 *Ladder diagram*

This tells us that we have A or the term B and C giving the output Q. Figure 4.45 shows the ladder diagram. Written in terms of Mitsubishi notation, the above expression might be:

$$X400 + X401{\cdot}X402 = Y430$$

In Siemens notation it might be:

$$I0.0 + I0.1{\cdot}I0.2 = Q2.0$$

As a further illustration, consider the Boolean expression:

$$A + \overline{B} = Q$$

Figure 4.46 shows the ladder diagram. Written in terms of Mitsubishi notation, the expression might be:

$$X400 + \overline{X401} = Y430$$

and in Siemens notation:

$$I0.0 + \overline{I0.1} = Q2.0$$

Consider the Exclusive-OR gate and its assembly from NOT, AND and OR gates, as shown in Figure 4.47.

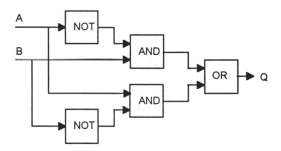

Figure 4.47 *XOR gate*

The input to the top AND gate is:

$$A \text{ and } \overline{B}$$

and so its output is:

$$A{\cdot}\overline{B}$$

The input to the lower AND gate is:

$\overline{A}$ and B

and so its output is:

$\overline{A} \cdot B$

Thus the Boolean expression for the output from the OR gate is:

$A \cdot \overline{B} + \overline{A} \cdot B = Q$

Consider a logic diagram with many inputs, as shown in Figure 4.48, and its representation by a Boolean expression and a ladder rung.

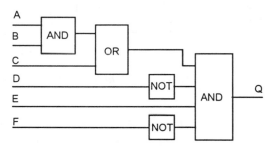

Figure 4.48 *Logic diagram*

For inputs A and B we obtain an output from the upper AND gate of A.B. From the OR gate we obtain an output of A.B + C. From the lower AND gate we obtain an output Q of:

$(A \cdot B + C) \cdot \overline{D} \cdot E \cdot \overline{F} = Q$

The ladder diagram to represent this is shown in Figure 4.49.

Figure 4.49 *Ladder diagram for Figure 4.48*

4.8 Function block diagrams

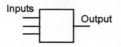

Figure 4.50 *Function block diagram*

(a)

(b)

Figure 4.51 *(a) Negated input, (b) negated output*

A format sometimes used for programmes is *function block diagrams*. These are basically the types of diagrams used above to describe logic systems, i.e. Figures 4.47 and 4.48. A function block is a program instruction unit which, when executed, yields one or more output values. Thus a block is represented in the manner shown in Figure 4.50 with the function name written in the box. Symbols are used for the function names. For example, the AND function is represented by &. The OR function by >=1, this is because there is an output if an input is greater than or equal to 1. A negated input is represented by a small circle on the input, a negative output by a small circle on the output (Figure 4.51).

To illustrate the form of such a diagram and its relationship to the ladder diagram, Figure 4.52 shows an OR gate. When A or B inputs are 1 then there is an output.

Figure 4.52 *Ladder diagram and equivalent functional block diagram*

Figure 4.53 shows a ladder diagram and its function block equivalent in Siemens notation. The = block is used to indicate an output from the system.

Figure 4.53 *Ladder diagram and equivalent function block diagram*

Figure 4.54 shows a ladder diagram involving the output having contacts acting as an input. The function block diagram equivalent can be shown as a feedback loop.

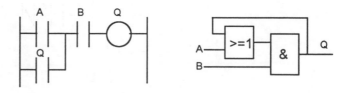

Figure 4.54 *Ladder diagram and equivalent function block diagram*

Consider the development of a function block diagram and ladder diagram for an application in which a pump is required to be activated and pump liquid into a tank when the start switch is closed, the level of liquid in a tank is below the required level and there is liquid in the reservoir from which it is to be pumped. What is required is an AND logic situation between the start switch input and a sensor input which is on when the liquid in the tank is below the required level. We might have a switch which is on until the liquid is at the required level. These two elements are then in an AND logic situation with a switch indicating that there is liquid in the reservoir. Suppose this switch gives an input when there is liquid. The function block diagram, and the equivalent ladder diagram, is then of the form shown in Figure 4.55.

Figure 4.55 *Pump application*

Function block diagrams can also be drawn involving such elements as counters and timers. See later chapters for a discussion of these elements.

4.9 IEC standards

As will have been apparent from this chapter, different manufacturers have their own ideas of how PLCs should be programmed. The IEC has, however, produced a standard IEC 1131 part 3 (1993), also given as British Standard BS EN 61131/3, on programming PLCs. This classifies programming methods into two categories: *textural language* and *graphical language*. As the names imply, textural language involves text and graphical language graphical images like ladders and block diagrams. Textural language has two methods: *instruction list* and *structured text*. Graphical has two methods: *ladder diagrams* and *function block diagrams*. Standards are laid down for the writing of programs in these four methods with examples being given.

In this chapter the methods introduced are instruction list, ladder diagrams and function block diagrams. In this book the main emphasis is on instruction list and ladder programming.

4.10 Programming examples

The following tasks are intended to illustrate the application of the programming techniques given in this chapter.

A signal lamp is required to be switched on if a pump is running and the pressure is satisfactory, or if the lamp test switch is closed. For the inputs from the pump and the pressure sensors we have an AND logic situation since both are required if there is to be an output from the lamp. We, however, have an OR logic situation with the test switch in that it is required to give an output of lamp on regardless of whether there is a signal from the AND system. The function block diagram, the ladder diagram and the instruction list are thus of the form shown in Figure 4.56. Note that with the ladder diagram and the instruction list we have to tell the PLC when it has reached the end of the program by the use of the END instruction.

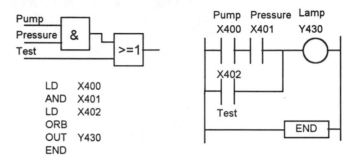

Figure 4.56 *Signal lamp task*

Consider another task. A motor is to be started when the start push button is pressed and remain on until the stop button is pressed. This requires the start button to be latched so that, after it has been pressed, the output remains on until the stop button is pressed. Figure 4.57 shows the function block diagram, the ladder diagram and the instruction list. Again note the use of the END instruction to complete the program.

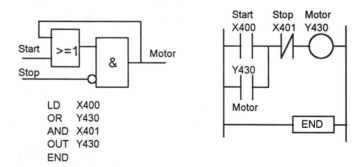

Figure 4.57 *Motor task*

As another example, consider a machine which has four sensors to detect when safety features are not active. When there is an input to any one of these sensors, the machine must stop and an alarm sound. Each sensor we can consider giving an input which has contacts that are normally closed so that the machine can run. When there is an input to the sensor the contacts open and the machine stops. With any one of the four sensors being required to switch the machine off we have an AND logic situation. Figure 4.58 shows the functional block diagram, the ladder diagram and the instruction list of a system that might be used.

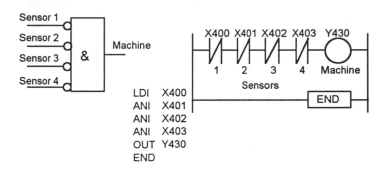

LDI X400
ANI X401
ANI X402
ANI X403
OUT Y430
END

Figure 4.59 *Machine safety controls task*

Problems

Questions 1 to 31 have four answer options: A, B, C or D. Choose the correct answer from the answer options.

1 Decide whether each of these statements is True (T) or False (F).

Figure 4.59 shows a ladder diagram rung for which:
(i) The input contacts are normally open.
(ii) There is an output when there is an input to the contacts.

A (i) T (ii) T
B (i) T (ii) F
C (i) F (ii) T
D (i) F (ii) T

Figure 4.59 *Problem 1*

2 Decide whether each of these statements is True (T) or False (F).

Figure 4.60 shows a ladder diagram rung for which:
(i) The input contacts are normally open.
(ii) There is an output when there is an input to the contacts.

A (i) T (ii) T
B (i) T (ii) F
C (i) F (ii) T
D (i) F (ii) T

Figure 4.60 *Problem 2*

3 Decide whether each of these statements is True (T) or False (F).

Figure 4.61 shows a ladder diagram rung for which:
(i) When only input 1 contacts are activated, there is an output.
(ii) When only input 2 contacts are activated, there is an output.

Inputs

Figure 4.61 *Problem 3*

A (i) T (ii) T
B (i) T (ii) F
C (i) F (ii) T
D (i) F (ii) T

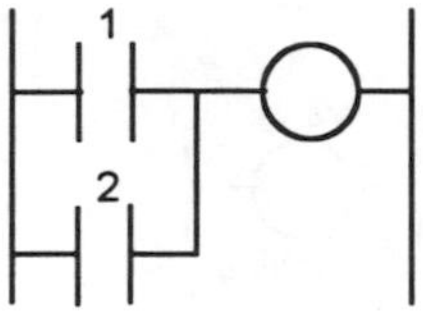

4 Decide whether each of these statements is True (T) or False (F).

Figure 4.62 shows a ladder diagram rung for which there is an output when:
(i) Inputs 1 and 2 are both activated.
(ii) Either one of inputs 1 and 2 is not activated.

Figure 4.62 *Problem 4*

A (i) T (ii) T
B (i) T (ii) F
C (i) F (ii) T
D (i) F (ii) T

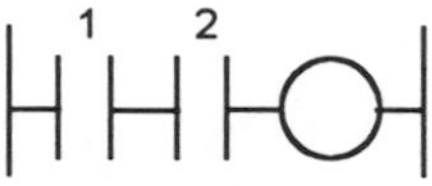

5 Decide whether each of these statements is True (T) or False (F).

Figure 4.63 shows a ladder diagram rung for which there is an output when:
(i) Inputs 1 and 2 are both activated.
(ii) Input 1 or 2 is activated.

Figure 4.63 *Problem 5*

A (i) T (ii) T
B (i) T (ii) F
C (i) F (ii) T
D (i) F (ii) T

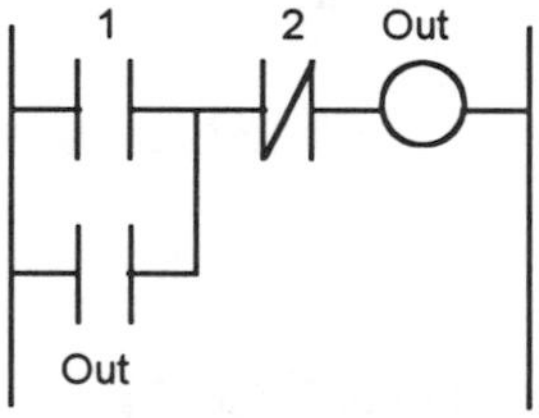

6 Decide whether each of these statements is True (T) or False (F).

Figure 4.64 shows a ladder diagram rung for which there is an output when:
(i) Input 1 is momentarily activated before reverting to its normally open state.
(ii) Input 2 is activated.

Figure 4.64 *Problem 6*

A (i) T (ii) T
B (i) T (ii) F
C (i) F (ii) T
D (i) F (ii) T

Questions 7 to 10 refer to the following logic gate systems:

A AND
B OR
C NOR
D NAND

7 Which form of logic gate system is given by a ladder diagram with a rung having two normally open sets of contacts in parallel?

8 Which form of logic gate system is given by a ladder diagram with a rung having two normally closed gates in parallel?

9 Which form of logic gate system is given by a ladder diagram with a rung having two normally closed gates in series?

10 Which form of logic gate system is given by a ladder diagram with a rung having two normally open gates in series?

11 Decide whether each of these statements is True (T) or False (F).

The instruction list:

```
LD        X401
AND       X402
OUT       Y430
```

describes a ladder diagram rung for which there is an output when:
(i) Input X401 is activated but X402 is not.
(ii) Input X401 and input X402 are both activated.

A (i) T (ii) T
B (i) T (ii) F
C (i) F (ii) T
D (i) F (ii) T

12 Decide whether each of these statements is True (T) or False (F).

The instruction list:

```
LD        X401
OR        X402
OUT       Y430
```

describes a ladder diagram rung for which there is an output when:
(i) Input X401 is activated but X402 is not.
(ii) Input X402 is activated but X401 is not.

A (i) T (ii) T
B (i) T (ii) F
C (i) F (ii) T
D (i) F (ii) T

13 Decide whether each of these statements is True (T) or False (F).

The instruction list:

```
LD        X401
ANI       X402
OUT       Y430
```

describes a ladder diagram rung for which there is an output when:
(i) Input X401 is activated but X402 is not.
(ii) Input X401 and input X402 are both activated.

A (i) T (ii) T
B (i) T (ii) F
C (i) F (ii) T
D (i) F (ii) T

14 Decide whether each of these statements is True (T) or False (F).

The instruction list:

LDI   X401
ANI   X402
OUT   Y430

describes a ladder diagram rung for which there is an output when:
(i) Input X401 is activated but X402 is not.
(ii) Input X401 and input X402 are both activated.

A (i) T (ii) T
B (i) T (ii) F
C (i) F (ii) T
D (i) F (ii) T

15 Decide whether each of these statements is True (T) or False (F).

The instruction list:

LD   X401
OR   Y430
ANI   X402
OUT   Y430

describes a ladder diagram rung for which there is:
(i) An output when input X401 is momentarily activated.
(ii) No output when X402 is activated.

A (i) T (ii) T
B (i) T (ii) F
C (i) F (ii) T
D (i) F (ii) T

16 Decide whether each of these statements is True (T) or False (F).

The instruction list:

A   I0.1
A   I0.2
=   Q2.0

describes a ladder diagram rung for which there is an output when:
(i) Input I0.1 is activated but I0.2 is not.
(ii) Input I0.1 and input I0.2 are both activated.

A (i) T (ii) T
B (i) T (ii) F
C (i) F (ii) T
D (i) F (ii) T

17 Decide whether each of these statements is True (T) or False (F).

The instruction list:

A   I0.1
O   I0.2
=   Q2.0

describes a ladder diagram rung for which there is an output when:
(i) Input I0.1 is activated but I0.2 is not.
(ii) Input I0.2 is activated but I0.1 is not.

A (i) T (ii) T
B (i) T (ii) F
C (i) F (ii) T
D (i) F (ii) T

18 Decide whether each of these statements is True (T) or False (F).

The instruction list:

A   I0.1
AN   I0.2
=   Q2.0

describes a ladder diagram rung for which there is an output when:
(i) Input I0.1 is activated but I0.2 is not.
(ii) Input I0.1 and input I0.2 are both activated.

A (i) T (ii) T
B (i) T (ii) F
C (i) F (ii) T
D (i) F (ii) T

19 Decide whether each of these statements is True (T) or False (F).

The instruction list:

AN   I0.1
AN   I0.2
=   Q2.0

describes a ladder diagram rung for which there is an output when:
(i) Input I0.1 is activated but I0.2 is not.
(ii) Input I0:1 and input I0.2 are both activated.

A (i) T (ii) T
B (i) T (ii) F
C (i) F (ii) T
D (i) F (ii) T

20 Decide whether each of these statements is True (T) or False (F).

The instruction list:

A   I0.1
O   Q2.0
AN   I0.2
=   Q2.0

describes a ladder diagram rung for which there is:
(i) An output when input I0.1 is momentarily activated.
(ii) No output when I0.2 is activated .

A (i) T (ii) T
B (i) T (ii) F
C (i) F (ii) T
D (i) F (ii) T

Problems 21 to 24 concern Boolean expressions for two inputs A and B.

A Input A is in series with input B, both inputs being normally off.
B Input A is in parallel with input B, both inputs being normally off.
C Input A, normally off, is in series with input B which is normally on.
D Input A is in parallel with input B, both inputs being normally on.

21 Which arrangement of inputs is described by the Boolean relationship A·B?

22 Which arrangement if inputs is described by the Boolean relationship A + B?

23 Which arrangement of inputs is described by the Boolean relationship $\overline{A} + \overline{B}$?

24 Which arrangement of inputs is described by the Boolean relationship A·$\overline{B}$?

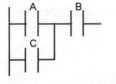

Figure 4.65 *Problem 25*

25 The arrangement of inputs in Figure 4.65 is described by the Boolean expression:

A A.B.C
B (A + C).B
C (A + B).C
D A·C + B

Figure 4.66 *Problem 26*

Figure 4.67 *Problem 27*

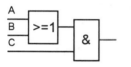

Figure 4.68 *Problem 28*

26 Decide whether each of these statements is True (T) or False (F).

For the function block diagram in Figure 4.66, there is an output:
(i) When A is 1.
(ii) When B is 1.

A (i) T (ii) T
B (i) T (ii) F
C (i) F (ii) T
D (i) F (ii) T

27 Decide whether each of these statements is True (T) or False (F).

For the function block diagram in Figure 4.67, there is an output:
(i) When A is 1.
(ii) When B is 1.

A (i) T (ii) T
B (i) T (ii) F
C (i) F (ii) T
D (i) F (ii) T

28 Decide whether each of these statements is True (T) or False (F).

For the functional block diagram in Figure 4.68, there is an output:
(i) When A is 1, B is 0 and C is 0.
(ii) When A is 0, B is 1 and C is 1.

A (i) T (ii) T
B (i) T (ii) F
C (i) F (ii) T
D (i) F (ii) T

29 Decide whether each of these statements is True (T) or False (F).

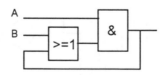

Figure 4.69 *Problem 29*

For the function block diagram in Figure 4.69, with A being a steady input condition and B a momentary input, there is an output:
(i) When A is 1 and B is 0.
(ii) When A is 0 and B is 1.

A (i) T (ii) T
B (i) T (ii) F
C (i) F (ii) T
D (i) F (ii) T

30 Figure 4.70(a) shows a ladder diagram. Which of the function block diagrams in Figure 4.70(b) is its equivalent?

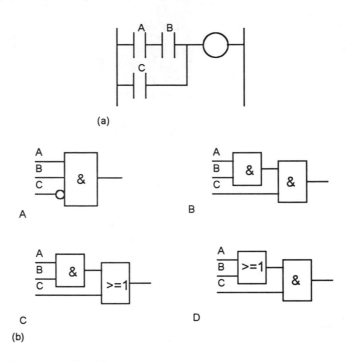

(a)

(b)

Figure 4.70 *Problem 30*

31 Figure 4.71(a) shows a function block diagram. Which of the ladder diagrams in Figure 4.71(b) is the equivalent?

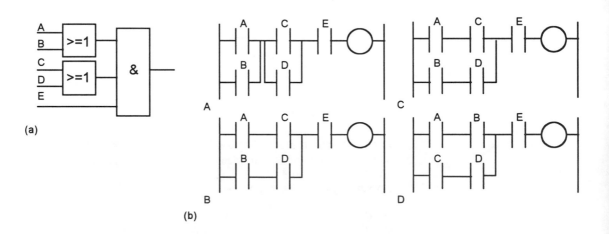

(a)

(b)

Figure 4.71 *Problem 31*

32 Draw the ladder rungs to represent:

(a) Two switches are normally open and both have to be closed for a motor to operate.

(b) Either of two, normally open, switches have to be closed for a coil to be energised and operate an actuator.

(c) A motor is switched on by pressing a spring-return push button start switch, and the motor remains on until another spring-return push button stop switch is pressed.

(d) A lamp is to be switched on if there is an input from sensor A or sensor B.

(e) A light is to come on if there is no input to a sensor.

(f) A solenoid valve is to be activated if sensor A gives an input.

5 Internal relays

This chapter continues on from Chapter 4 and introduces *internal relays*. A variety of other terms are often used to describe these elements, e.g. *auxiliary relays, markers, flags, coils, bit storage*. These are one of the elements giving special built-in functions with PLCs and are very widely used in programming. A small PLC might have a hundred or more internal relays, some of them being battery backed so that they can be used in situations where it is necessary to ensure safe shutdown of plant in the event of power failure. Later chapters consider other common built-in elements.

5.1 Internal relays   In PLCs there are elements that are used to hold data, i.e. bits, and behave like relays, being able to be switched on or off and switch other devices on or off. Hence the term *internal relay*. Such internal relays do not exist as real-world switching devices but are merely bits in the storage memory that behave in the same way as relays. For programming, they can be treated in the same way as an external relay output and input. Thus inputs to external switches can be used to give an output from an internal relay. This then results in the internal relay contacts being used, in conjunction with other external input switches to give an output, e.g. activate a motor. Thus we might have (Figure 5.1):

> On one rung of the program:
> > Inputs to external inputs activate the internal relay output.

> On a later rung of the program:
> > As a consequence of the internal relay output:
> > internal relay contacts are activated and so control some output.

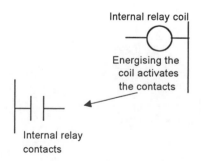

Figure 5.1 *Internal relay*

In using an internal relay, it has to be activated on one rung of a program and then its output used to operate switching contacts on another rung, or rungs, of the program. Internal relays can be programmed with as many sets of associated contacts as desired.

To distinguish internal relay outputs from external relay outputs, they are given different types of addresses. Different manufacturers tend to use different terms for internal relays and different ways of expressing their addresses. For example, Mitsubishi uses the term *auxiliary relay* or *marker* and the notation M100, M101, etc. Siemens uses the term *flag* and notation F0.0, F0.1, etc. Sprecher+Schuh uses the term *coil* and notation C001, C002, etc. Telemecanique uses the term *bit* and notation B0, B1, etc. Toshiba uses the term *internal relay* and notation R000, R001, etc. Allen Bradley uses the term *bit storage* and notation in the PLC-5 of the form B3/001, B3/002, etc.

5.2 Internal relays in programs

With ladder programs, an internal relay output is represented using the symbol for an output device, namely ( ) or O, with an address which indicates that it is an internal relay rather than an external relay. Thus, with a Mitsubishi PLC, we might have the address M100, the M indicating that it is an internal relay or marker rather than an external device. The internal relay switching contacts are designated with the symbol for an input device, namely ‖, and given the same address as the internal relay output, e.g. M100.

5.2.1 Programs with multiple input conditions

As an illustration of the use that can be made of internal relays, consider the following situation. A system is to be activated when two different sets of input conditions are realised. We might just program this as an AND logic gate system; however, if a number of inputs have to be checked in order that each of the input conditions can be realised, it may be simpler to use an internal relay. The first input conditions then are used to give an output to an internal relay. This has associated contacts which then become part of the input conditions with the second input.

Figure 5.2 shows a ladder program for such a task. For the first rung: when input 1 or input 3 is closed and input 2 closed, then internal relay IR 1 is activated. This results in the contacts IR 1 closing. If input 4 is then activated, there is an output from output 1. Such a task might be involved in the automatic lifting of a barrier when someone approaches from either side. Input 1 and input 3 are inputs from photoelectric sensors that detect the presence of a person, approaching or leaving from either side of the barrier, input 1 being activated from one side of it and input 3 from the other. Input 2 is an enabling switch to enable the system to be closed down. Thus when input 1 or input 3, and input 2, are activated, there is an output from the internal relay 1. This will close the internal relay contacts. If input 4, perhaps a limit switch, detects that the barrier is closed then it is activated and closes. The result is then an output from Out 1, a motor

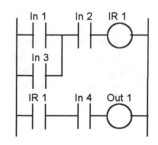

Figure 5.2 *Internal relay*

which lifts the barrier. If the limit switch detects that the barrier is already open, the person having passed through it, then it opens and so output 1 is no longer energised and a counterweight might then close the barrier. The internal relay has enabled two parts of the program to be linked, one part being the detection of the presence of a person and the second part the detection of whether the barrier is already up or down.

Figure 5.3(a) shows how Figure 5.2 would appear in Mitsubishi notation and Figure 5.3(b) in Siemens notation.

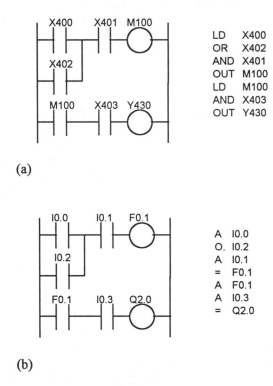

(a)

(b)

Figure 5.3 *(a) Mitsubishi notation, (b) Siemens notation*

Figure 5.4 is another example of this type of ladder program. The output 1 is controlled by two input arrangements. The first rung shows the internal relay IR 1 which is energised if the input In 1 or In 2 is activated and closed. The second rung shows the internal relay IR 2 which is energised if the inputs In 3 and In 4 are both energised. The third rung shows that the output Out 1 is energised if the internal relay IR 1 or IR 2 is activated. Thus there is an output from the system if either of two sets of input conditions is realised.

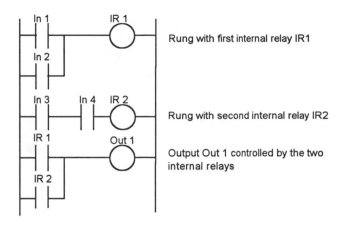

Figure 5.4 *Use of two internal relays*

5.2.2 Latching programs

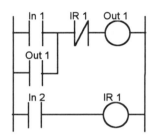

Figure 5.5 *Resetting latch*

Another use of internal relays is for resetting a latch circuit. **Figure 5.5** shows an example of such a ladder program. When the input 1 contacts are momentarily closed, there is an output at Out 1. This closes the contacts for Out 1 and so maintains the output, even when input 1 opens. When input 2 is closed, the internal relay IR 1 is energised and so opens the IR 1 contacts, which are normally closed. Thus the output Out 1 is switched off and so the output is unlatched. Figure 5.6 shows Figure 5.5 in (a) Mitsubishi, (b) Siemens, (c) Telemecanique notations.

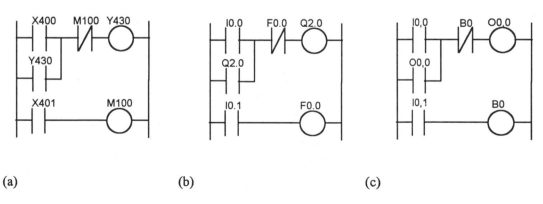

(a)   (b)   (c)

Figure 5.6 *(a) Mitsubishi, (b) Siemens, (c) Telemecanique notations*

Consider a situation requiring latch circuits where there is an automatic machine that can be started or stopped using pushbutton switches. A latch circuit is used to start and stop the power being applied to the machine. The machine has several outputs which can be turned on if the power has been turned on and are off if the power is off. It would be possible to devise a ladder diagram which has individually latched controls for each such output. However, a simpler method is to use an internal relay.

Figure 5.7 shows such a ladder diagram. The first rung has the latch for keeping the internal relay IR 1 on when the start switch gives a momentary input. The second rung will then switch the power on. The third rung will also switch on and give output Out 2 if input 2 contacts are closed. The third rung will also switch on and give output Out 3 if input 3 contacts are closed. Thus all the outputs can be switched on when the start push button is activated. All the outputs will be switched off if the stop switch is opened. Thus all the outputs are latched by IR 1.

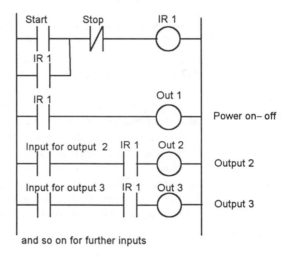

Figure 5.7 *Starting of multiple outputs*

5.3 Battery-backed relays

If the power supply is cut off from a PLC while is it being used, all the output relays and internal relays will be turned off. Thus when the power is restored, all the contacts associated with those relays will be set differently from when the power was on. Thus, if the PLC was in the middle of some sequence of control actions, it would resume at a different point in the sequence. To overcome this problem, some internal relays have battery back-up so that they can be used in circuits to ensure a safe shutdown of plant in the event of a power failure and so enable it to restart in an appropriate manner. Such battery-backed relays retain their state of activation, even when the power supply is off. The relay is said to have been made *retentive*.

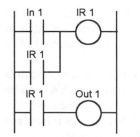

Figure 5.8 *Battery-backed relay program*

As an example of the use of such a relay, Figure 5.8 shows a ladder diagram for a system designed to cope with a power failure. IR 1 is a battery-backed internal relay. When input 1 contacts close, output IR 1 is energised. This closes the IR 1 contacts, latching so that IR 1 remains on even if input 1 opens. The result is an output from Out 1. If there is a power failure, IR 1 still remains energised and so the IR 1 contacts remain closed and there is an output from Out 1.

With Mitsubishi PLCs, battery-backed internal relay circuits use M300 to M377 as addresses for such relays. Other manufacturers use different addresses.

5.4 One-shot operation

One of the functions provided by some PLC manufacturers is the ability to program an internal relay so that its contacts are activated for just one cycle, i.e. one scan through the ladder program. Hence it provides a fixed duration pulse at its contacts when operated. This function is often termed *one-shot*.

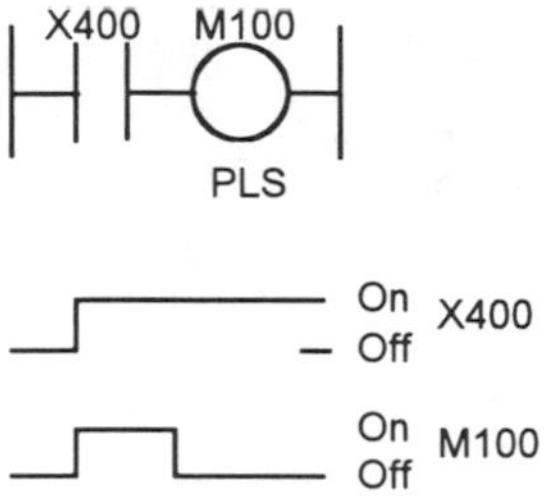

Figure 5.9 illustrates this with a simple ladder diagram for a Mitsubishi PLC. When X400 contacts close, the output internal relay M100 is activated. Under normal circumstances, M100 would remain on for as long as the X400 contacts were closed. However, if M100 has been programmed for pulse operation, M100 only remains on for a fixed period of time, one program cycle. It then goes off, regardless of X400 being on. The programming instructions that would be used are:

```
LD      X400
PLS     M100
```

Figure 5.9 *Pulse operation*

The above represents pulse operation when the input goes from off to on, i.e. is positive-going. If, in Figure 5.9, X400 is made normally closed, rather than normally open, then the pulse occurs when the input goes from on to off, i.e. is negative-going.

Other manufacturers have different forms for such a ladder program and use different programming instructions. For example, Sprecher+Schuh would have the ladder diagram shown in Figure 5.10 and the following programming instructions:

```
STR     X001
F-05    DION
OUT     C001
```

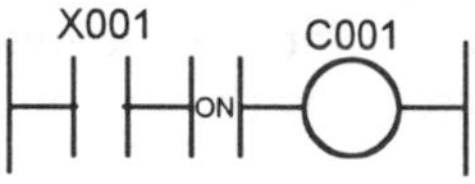

Figure 5.10 *Pulse operation*

The IEC symbol for a pair of pulse contacts is either |P| for a positive-going signal or |N| for a negative going signal. In the above, a ready-made pulse function was available. Some manufacturers, however, use basic logic elements to build up such a function. See section 5.5 for examples.

Such a pulse providing relay, or built-up function, is used to create pulses for the resetting of counters and timers and marking the start of cycles.

5.5 Set and reset

Another function which is often available is the ability to set and reset an internal relay. The set instruction causes the relay to self-hold, i.e. latch. It then remains in that condition until the reset instruction is received. The term *flip-flop* is often used.

Figure 5.11 shows an example of a ladder diagram involving such a function. Activation of the first input, X400, causes the output Y430 to be turned on and set, i.e. latched. Thus if the first input is turned off, the output remains on. Activation of the second input, X401, causes the output Y430 to be reset, i.e. turned off and latched off. Thus the output Y430 is on for the time between X400 being momentarily switched on and X401 being momentarily switched on. Between the two rungs indicated for the set and reset operations, there could be other rungs for other activities to be carried out, the set rung switching on an output at the beginning of the sequence and off at the end.

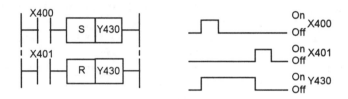

Figure 5.11 *Set and reset*

The programming instructions for the ladder rungs are:

LD	X400
S	Y430

Other program rungs

LD	X401
R	Y430

Figure 5.12 shows the equivalent ladder diagram for the set-reset function with a Siemens PLC, the programming instructions (F being used to indicate an internal relay) then being:

A	I0.0
S	F0.0
A	I0.1
R	F0.0
A	F0.0
=	Q2.0

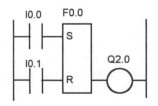

Figure 5.12 *Set and reset*

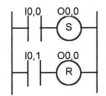

Figure 5.13 *Set and reset*

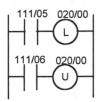

Figure 5.14 *Latch and unlatch*

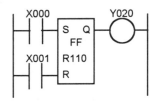

Figure 5.15 *Flip-flop*

With a Telemecanique PLC the ladder diagram would be as shown in Figure 5.13 and the programming instructions would be:

L	I0,0
S	O0,0
L	I0,1
R	O0,0

With an Allen Bradley PLC, the term latch and unlatch is used. Figure 5.14 shows the ladder diagram. Toshiba uses the term flip-flop and Figure 5.15 shows the ladder diagram.

Figure 5.16 shows how the set-reset function can be used to build the pulse (one-shot) function described in the previous section. Figure 5.16(a) shows it for a Siemens PLC (F indicates internal relay) and Figure 5.16(b) for a Telemecanique PLC (B indicates internal relay). The mode of operation is the same for each. An input (I0.0, I0,0) causes the internal relay (B0, F0.0) in the first rung to be activated. This results, second rung, in the set–reset internal relay being set. This setting action results in the internal relay (F0.1, B1) in the first rung opening and so, despite there being an input in the first rung, the internal relay (BO, F0.0) opens. However, because the rungs are scanned in sequence from top to bottom, a full cycle must elapse before the internal relay in the first rung opens. A pulse of duration one cycle has thus been produced. The system is reset when the input (I0.0, I0,1) ceases.

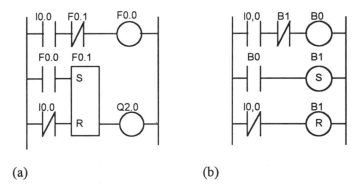

(a)   (b)

Figure 5.16 *Pulse function, (a) Siemens PLC, (b) Telemecanique PLC*

5.6 Master control relay

When large numbers of outputs have to be controlled, it is sometimes necessary for whole sections of ladder diagrams to be turned on or off when certain criteria are realised. This could be achieved by including the contacts of the same internal relay in each of the rungs so that its operation affects all of them. An alternative is to use a *master control relay*.

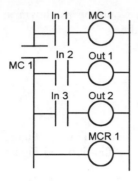

Figure 5.17 *Use of a master control relay*

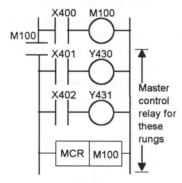

Figure 5.18 *Use of a master control relay*

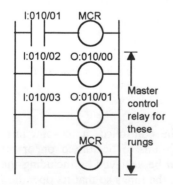

Figure 5.19 *Use of a master control relay*

Figure 5.17 illustrates the use of such a relay to control a section of a ladder program. With no input to input 1, the output internal relay MC 1 is not energised and so its contacts are open. This means that all the rungs between where it is designated to operate and the rung on which its reset MCR or another master control relay is located are switched off. Assuming it is designated to operate from its own rung, then we can imagine it to be located in the power line in the position shown and so rungs 2 and 3 are off. When input 1 contacts close, the master relay MC 1 is energised. When this happens, all the rungs between it and the rung with its reset MCR 1 are switched on. Thus outputs 1 and 2 cannot be switched on by inputs 2 and 3 until the master control relay has been switched on. The master control relay 1 acts only over the region between the rung it is designated to operate from and the rung on which MCR 1 is located.

With a Mitsubishi PLC, an internal relay can be designated as a master control relay by programming it accordingly. Thus to program an internal relay M100 to act as master control relay contacts the program instruction is:

    MC   M100

To program the resetting of that relay, the program instruction is:

    MCR   M100

Thus for the ladder diagram shown in Figure 5.18, being Figure 5.17 with Mitsubishi addresses, the program instructions are:

    LD   X400
    OUT   M100
    MC   M100
    LD   X401
    OUT   Y430
    LD   X402
    OUT   Y431
    MC   M100

Figure 5.19 shows the format used by Allen Bradley. To end the control of one master control relay (MCR), a second master control relay (MCR) is used with no contacts or logic preceding it. It is said to be programmed unconditionally.

A program might use a number of master control relays, enabling various sections of a ladder program to be switched in or out. Figure 5.20 shows a ladder program in Mitsubishi format involving two master control relays. With M100 switched on, but M101 off, the sequence is: rungs 1, 3, 4, 6, etc. The end of the M100 controlled section is indicated by the occurrence of the other master control relay, M101. With M101 switched on, but M100 off, the sequence is: rungs 2, 4, 5, 6, etc. The end of this section is indicated by the presence of the reset. This reset has to be used since the rung is not followed immediately by another master control relay.

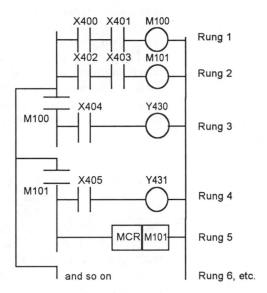

Figure 5.20 *Example showing more than one master control relay*

The program instruction list for Figure 5.20 is:

    LD   X400
    AND   X401
    OUT   M100
    LD   X402
    AND   X403
    OUT   M101
    MC   M100
    LD   X404
    OUT   Y430
    MC   M101
    LD   X405
    OUT   Y431
    MCR   M101
    and so on.

Such an arrangement could be used to switch on one set of ladder rungs if one type of input occurs, and another set of ladder rungs if a different input occurs.

5.7 Jump

A function often provided with PLCs is the *conditional jump*. If the appropriate conditions are met, this function enables part of a ladder program to be jumped over. Figure 5.21 illustrates this in a general manner. When there is an input to In 1, its contacts close and there is an output to the jump relay. This then results in the program jumping to the rung in which the jump end occurs, so missing out intermediate program rungs. Thus, in this case, when there is an input to In 1, the program jumps to rung 4 and then proceeds with rungs 5, 6, etc. When there is no input to In 1, the jump relay is not energised and the program then proceeds to rungs 2, 3, etc.

Such a facility enables programs to be designed such that:

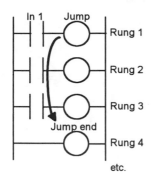

> If certain conditions are met then certain events occur, if they are not met then other events occur.

Figure 5.21 *Jump*

Thus, for example, we might need to design a system so that if the temperature is above 60°C a fan is switched on, and if below that temperature no action occurs.

Figure 5.22 shows the above ladder program in the form used by Mitsubishi. The jump instruction is denoted by CJP (conditional jump) and the place to which the jump occurs is denoted by EJP (end of jump). The condition that the jump will occur is then that there is an input to X400. When that happens the rungs involving inputs X401 and X403 are ignored and the program jumps to continue with the rungs following the end jump instruction with the same number as the start jump instruction, i.e. in this case EJP 700.

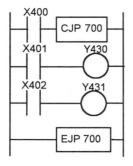

Figure 5.22 *Jump*

With the Allen Bradley PLC-5 format the jump takes place from the jump instruction (JMP) to the label instruction (LBL). The JMP instruction is given a three-digit number from 000 to 255 and the LBL instruction the same number. Figure 5.23 shows a ladder program in this format.

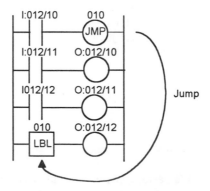

Figure 5.23 *Jump*

Jumps within jumps are possible. For example, we might have the situation shown in Figure 5.24. If the condition for the jump instruction 1 is realised then the program jumps to rung 8. If the condition is not met then the program continues to rung 3. If the condition for the jump instruction 2 is realised then the program jumps to rung 6. If the condition is not met then the program continues through the rungs.

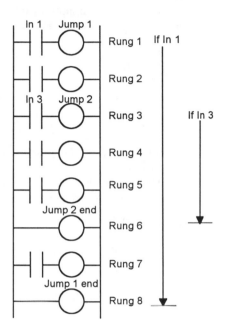

Figure 5.24 *Jumps within jumps*

Thus if we have an input to In 1, the rung sequence is rung 1, 8, etc. If we have no input to In 1 but an input to In 3, then the rung sequence is 1, 2, 6, 7, 8, etc. If we have no input to In 1 and no input to In 3, the rung sequence is 1, 2, 3, 4, 5, 6, 7, 8, etc. The jump instruction enables different groups of program rungs to be selected, depending on the conditions occurring.

Problems   *Questions 1 to 21 have four answer options: A, B, C or D. Choose the correct answer from the answer options.*

Problems 1 to 3 refer to Figure 5.25 which shows a ladder diagram with an internal relay, designated IR 1, two inputs In 1 and In 2, and an output Output 1.

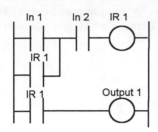

Figure 5.25 *Problems 1 and 2*

1 Decide whether each of these statements is True (T) or False (F).

For the ladder diagram shown in Figure 5.25, there is no output from output 1 when:
(i)   There is just an input to In 1.
(ii) There is just an input to In 2.

A (i) T (ii) T
B (i) T (ii) F
C (i) F (ii) T
D (i) F (ii) F

2 Decide whether each of these statements is True (T) or False (F).

For the ladder diagram shown in Figure 5.25, there is no output from output 1 when:
(i)   There is an input to In 2 and a momentary input to In 1.
(ii) There is an input to In 1 or an input to In 2.

A (i) T (ii) T
B (i) T (ii) F
C (i) F (ii) T
D (i) F (ii) F

3 Decide whether each of these statements is True (T) or False (F).

For the ladder diagram shown in Figure 5.25, the internal relay:
(i)   Switches on when there is just an input to In 1.
(ii) Switches on when there is an input to In 1 and to In 2.

A (i) T (ii) T
B (i) T (ii) F
C (i) F (ii) T
D (i) F (ii) F

Problems 4 to 6 refer to Figure 5.26 which shows a ladder diagram involving internal relays IR 1 and IR 2, inputs In 1, In 2 and In 3, and output Output 1.

4 Decide whether each of these statements is True (T) or False (F).

For the ladder diagram shown in Figure 5.26, the internal relay IR 1 is energised when:
(i)   There is an input to In 1.
(ii) There is an input to In 3.

A (i) T (ii) T
B (i) T (ii) F
C (i) F (ii) T
D (i) F (ii) F

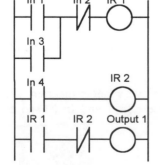

Figure 5.26 *Problems 4 to 6*

5 Decide whether each of these statements is True (T) or False (F).

For the ladder diagram shown in Figure 5.26, the internal relay IR 2 is energised when:
(i) Internal relay IR 1 is energised.
(ii) Input 4 is energised.

A (i) T (ii) T
B (i) T (ii) F
C (i) F (ii) T
D (i) F (ii) F

6 Decide whether each of these statements is True (T) or False (F).

For the ladder diagram shown in Figure 5.26, there is an output from Output 1 when:
(i) There are inputs to just In 1, In 2 and In 4.
(ii) There are inputs to just In 3 and In 4.

A (i) T (ii) T
B (i) T (ii) F
C (i) F (ii) T
D (i) F (ii) F

Questions 7 and 8 refer to Figure 5.27 which shows a ladder diagram involving a battery-backed relay IR1, two inputs In 1 and In 2 and an output Output 1.

7 Decide whether each of these statements is True (T) or False (F).

For the ladder diagram shown in Figure 5.27, there is an output from Output 1 when:
(i) There is a short duration input to In 1.
(ii) There is no input to In 2.

A (i) T (ii) T
B (i) T (ii) F
C (i) F (ii) T
D (i) F (ii) F

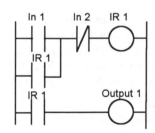

Figure 5.27 *Problems 7 and 8*

8 Decide whether each of these statements is True (T) or False (F).

For the ladder diagram shown in Figure 5.27:
(i) The input In 1 is latched by the internal relay so that the internal relay IR 1 remains energised, even when the input In 1 ceases.
(ii) Because the internal relay IR 1 is battery-backed, once there is an output from Output 1, it will continue, even when the power is switched off, until there is an input to In 2.

A (i) T (ii) T
B (i) T (ii) F
C (i) F (ii) T
D (i) F (ii) F

9 When the program instructions LD X100, PLS M400 are used for a ladder rung, the internal relay M400 will:

A Remain on even when the input to X100 ceases.
B Remain closed unless there is a pulse input to X100.
C Remain on for one program cycle when there is an input to X100.
D Remain closed for one program cycle after there is an input to X100.

10 When the program instructions LDI X100, PLS M400 are used for a ladder rung, the internal relay M400 will:

A Remain on when the input to X100 ceases.
B Remain on when there is a pulse input to X100.
C Remain on for one program cycle when there is an input to X100.
D Remain on for one program cycle after the input to X100 ceases.

11 A Mitsubishi ladder program has the program instructions LD X100, S M200, LD X101, R M200, followed by other instructions for further rungs. There is the sequence: an input to the input X100, the input to X100 ceases, some time elapses, an input to the input X101, the input to X101 ceases, followed by inputs to later rungs. The internal relay M200 will remain on:

A For one program cycle from the start of the input to X100.
B From the start of the input to X100 to the start of the input to X101.
C From the start of the input to X100 to the end of the input to X101.
D From the end of the input to X100 to the end of the input to X101.

12 A Siemens ladder program has the program instructions A I0.0, S F0.0, A I0.1, R F0.0, A F0.0, = Q2.0, followed by other instructions for further rungs. There is the sequence: an input to the input I0.0, the input to I0.0 ceases, some time elapses, an input to the input I0.1, the input to I0.1 ceases, followed by inputs to later rungs. The internal relay F0.0 will remain on:

A For one program cycle from the start of the input to I0.0.
B From the start of the input to I0.0 to the start of the input to I0.1.
C From the start of the input to I0.0 to the end of the input to I0.1.
D From the end of the input to I0.0 to the end of the input to I0.1.

13 A Telemecanique ladder program has the program instructions L I0,0, S O0,0, L I0,1, R O0,0, followed by other instructions for further rungs. There is the sequence: an input to the input I0,0, the input to I0,0 ceases, some time elapses, an input to the input I0,1, the input to I0,1 ceases, followed by inputs to later rungs. The internal relay O0,0 will remain on:

A For one program cycle from the start of the input to I0,0.
B From the start of the input to I0,0 to the start of the input to I0,1.
C From the start of the input to I0,0 to the end of the input to I0,1.
D From the end of the input to I0,0 to the end of the input to I0,1.

14 An output is required from output Y430 which lasts for one cycle after an input to X100 starts. This can be given by a ladder program with the instructions:

A LD X100, Y430
B LD X100, M100, LD M100, Y 430
C LD X100, PLS M100, LD M100, Y 430
D LD X400, PLS M100, LDI M100, Y430

Questions 15 and 16 refer to Figure 5.28, which are two versions of the same ladder diagram according to two different PLC manufacturers. In (a) which uses Siemens notation, I is used for inputs, F for internal relays and Q for the output. In (b) which uses Telemecanique notation, I is used for inputs and B for internal relays.

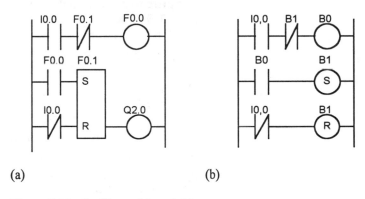

(a)   (b)

Figure 5.28 *Problems 15 and 16*

15 For the ladder diagram shown in Figure 5.28(a), when there is an input to I0.0, the output Q2.0:

A Comes on and remains on for one cycle.
B Come on and remains on.
C Goes off and remains off for one cycle.
D Goes off and remains off.

16 For the ladder diagram shown in Figure 5.28(b), when there is an input to I0,0, the internal relay B1:

A Comes on and remains on for one cycle.
B Come on and remains on.
C Goes off and remains off for one cycle.
D Goes off and remains off.

Questions 17 and 18 refer to Figure 5.29 which shows a Toshiba ladder program with inputs X000, X001 and X002, an output Y020 and a flip-flop R110.

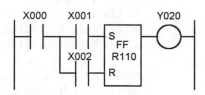

Figure 5.29 *Problems 17 and 18*

17 Decide whether each of these statements is True (T) or False (F).

For there to be an output from Y020 there must be inputs to:
(i) X000.
(ii) X001.

A (i) T (ii) T
B (i) T (ii) F
C (i) F (ii) T
D (i) F (ii) F

18 Decide whether each of these statements is True (T) or False (F).

With an input to X001, then:
(i) An input to X001 causes the output to come on.
(ii) An input to X002 causes the output to come on.

A (i) T (ii) T
B (i) T (ii) F
C (i) F (ii) T
D (i) F (ii) F

19 Decide whether each of these statements is True (T) or False (F).

A master control relay can be used to:
(i) Turn on a section of a program when certain criteria are met.
(ii) Turn off a section of a program when certain criteria are not met.

A (i) T (ii) T
B (i) T (ii) F
C (i) F (ii) T
D (i) F (ii) F

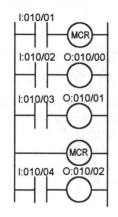

Figure 5.30 *Problems 20 and 21*

Questions 20 and 21 refer to Figure 5.30 which shows a ladder program in Allen Bradley format.

20 Decide whether each of these statements is True (T) or False (F).

When there is an input to I:010/01:
(i) An input to I:010/02 gives an output from O:010/00.
(ii) An input to I:010/03 gives an output from O:010/01.

A (i) T (ii) T
B (i) T (ii) F
C (i) F (ii) T
D (i) F (ii) F

21 Decide whether each of these statements is True (T) or False (F).

When there is no input to I:010/01:
(i) An input to I:010/02 gives no output from O:010/00.
(ii) An input to I:010/04 gives no output from O:010/02.

A (i) T (ii) T
B (i) T (ii) F
C (i) F (ii) T
D (i) F (ii) F

22 Devise ladder programs which can be used to:
(a) Maintain an output on, even when the input ceases and when there is a power failure.
(b) Switch on an output for a time of one cycle following a brief input.
(c) Switch on the power to a set of rungs.

6 Timers

In many control tasks there is a need to control time. For example, a motor or a pump might need to be controlled to operate for a particular interval of time, or perhaps be switched on after some time interval. PLCs thus have timers as built-in devices. Timers count fractions of seconds or seconds using the internal CPU clock (see Figure 1.8). This chapter shows how such timers can be programmed to carry out control tasks.

6.1 Types of timers

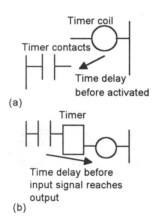

Figure 6.1 *Treatment of timers*

PLC manufacturers differ on how timers should be programmed and hence how they can be considered. A common approach is to consider timers to behave like relays with coils which when energised result in the closure or opening of contacts after some preset time. The timer is thus treated as an output for a rung with control being exercised over pairs of contacts elsewhere (Figure 6.1(a)). This is the predominant approach used in this book. Some treat a timer as a delay block which when inserted in a rung delays signals in that rung reaching the output (Figure 6.1(b)).

There are a number of different forms of timers that can be found with PLCs. With small PLCs there is likely to be just one form, the *on-delay timers*. These are timers which come on after a particular time delay (Figure 6.2(a)). *Off-delay timers* are on for a fixed period of time before turning off (Figure 6.2(b)). Another type of timer that occurs is the *pulse timer*. This timer switches on or off for a fixed period of time (Figure 6.2(c)).

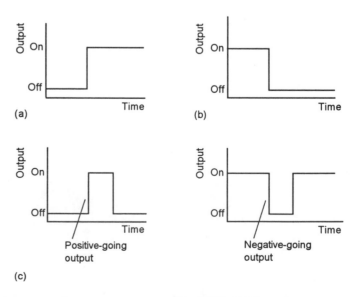

Figure 6.2 *Timers: (a) on-delay, (b) off-delay, (c) pulse*

With function block diagrams (see section 4.8) and the symbols used for timers by some manufacturers, TON is used to denote on-delay, TOF off-delay. On delay is also represented by T–0 and off-delay by 0–T. Pulse timers are denoted by TP.

The time duration for which a timer has been set is termed the preset and is set in multiples of the time base used. Some times bases are typically 10 ms, 100 ms, 1 s, 10 s and 100 s. Thus a preset value of 5 with a time base of 100 ms is a time of 500 ms. For convenience, where timers are involved in this text, a time base of 1 s has been used.

6.2 Programming timers

All PLCs generally have delay-on timers, small PLCs possibly having only this type of timer. Figure 6.3 shows a ladder rung diagram involving a delay-on timer. The timer is like a relay with a coil which is energised when the input In 1 occurs (rung 1). It then closes, after some preset time delay, its contacts on rung 2. Thus the output from Out 1 occurs some preset time after the input In 1 occurs.

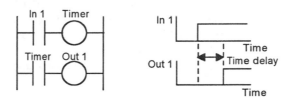

Figure 6.3 *Ladder diagram involving a delay-on timer*

Figure 6.4 shows how the above diagram, and program instructions, would appear when drawn in the form used by two different manufacturers of PLCs. The Mitsubishi figure (Figure 6.4(a)) considers the timer as an output which gives a delayed time reaction to contacts. With the Siemens figure (Figure 6.4(b), the manufacturer considers the timer to be a delay item in a rung, rather than as a relay. The symbol in the Siemens rectangle indicates a delay-on timer, the 0 coming after the T indicating the delay on. The time delay has been chosen to be 5 s. Techniques for the entry of preset time values vary. Often it requires the entry of a constant K command followed by the time interval in multiples of the time base used.

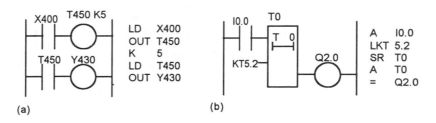

Figure 6.4 *(a) Mitsubishi, (b) Siemens*

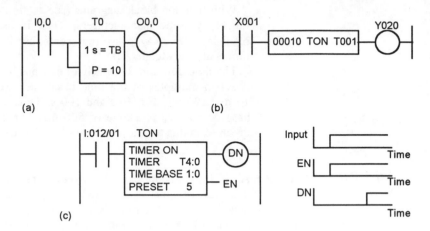

(a)   (b)

(c)

Figure 6.5 *(a) Telemecanique, (b) Toshiba, (c) Allen Bradley*

Figure 6.5 shows ladder diagrams for three other manufacturers, (a) Telemecanique, (b) Toshiba and (c) Allen Bradley. With the Allen Bradley diagram the DN (for done) signal is the signal that is produced when the timer has finished its action, the EN (for enable) is the signal which is a replica of the timer input and used as instantaneous contacts.

6.2.1 Sequencing

As an illustration of the use of a timer, consider the ladder diagram shown in Figure 6.6(a). When the input In 1 is on, the output Out 1 is switched on. The contacts associated with this output then start the timer. The contacts of the timer will close after the preset time delay, in this case 5.5 s. When this happens, output Out 2 is switched on. Thus, following the input In 1, Out 1 is switched on and followed 5.5 s later by Out 2. This illustrates how timed sequence of outputs can be achieved. Figure 6.6(b) shows the same operation where the format used by the PLC manufacturer is for the timer to institute a signal delay.

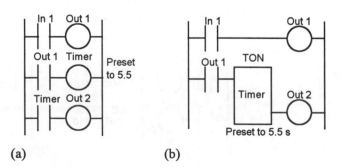

(a)   (b)

Figure 6.6 *Sequenced outputs*

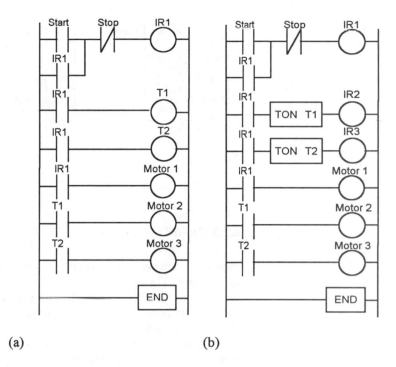

(a)   (b)

Figure 6.7 *Motor sequence*

Figure 6.7 shows two versions of how timers can be used to start three outputs, e.g. three motors, in sequence following a single start button being pressed. In (a) the timers are programmed as coils, whereas in (b) they are programmed as delays. When the start push button is pressed there is an output from internal relay IR1. This latches the start input. It also starts both the timers, T1 and T2, and motor 1. When the preset time for timer 1 has elapsed then its contacts close and motor 2 starts. When the preset time for timer 2 has elapsed then its contacts close and motor 3 starts. The three motors are all stopped by pressing the stop push button. Since this is seen as a complete program, the end instruction has been used.

6.2.2 Cascaded timers

Timers can be linked together, the term *cascaded* is used, to give longer delay times than are possible with just one timer. Figure 6.8(a) shows the ladder diagram for such an arrangement. Thus we might have timer 1 with a delay time of 999 s. This timer is started when there is an input to In 1. When the 999 s time is up, the contacts for timer 1 close. This then starts timer 2. This has a delay of 100 s. When this time is up, the timer 2 contacts close and there is an output from Out 1. Thus the output occurs 1099 s after the input to In 1. Figure 6.8(b) shows the Mitsubishi version of this ladder diagram and the program instructions for that ladder.

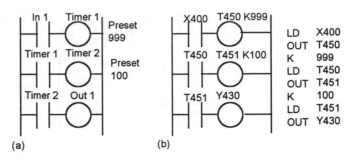

Figure 6.8 *Cascaded timers*

6.2.3 On–off cycle timer

Figure 6.9 shows how on-delay timers can be used to produce an *on–off cycle timer*. The timer is designed to switch on an output for 5 s, then off for 5 s, then on for 5 s, then off for 5 s, and so on. When there is an input to In 1 and its contacts close, timer 1 starts. Timer 1 is set for a delay of 5 s. After 5 s, it switches on timer 2 and the output Out 1. Timer 2 has a delay of 5 s. After 5 s, the contacts for timer 2, which are normally closed, open. This results in timer 1, in the first rung, being switched off. This then causes its contacts in the second rung to open and switch off timer 2. This results in the timer 2 contacts resuming their normally closed state and so the input to In 1 causes the cycle to start all over again.

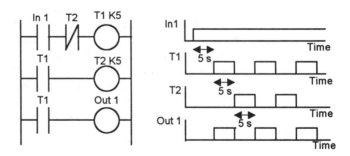

Figure 6.9 *On–off cycle timer*

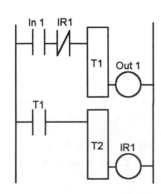

Figure 6.10 *On-off cycle timer*

Figure 6.10 shows how the above ladder program would appear in the format used with a timer considered as a delay, rather than as a coil. This might, for example, be with Siemans or Toshiba. When input In 1 closes, the timer T1 starts. After its preset time, there is an output to Out 1 and timer T2 starts. After its preset time there is an output to the internal relay IR1. This opens its contacts and stops the output from Out 1. This then switches off timer T2. The entire cycle can then repeat itself.

6.3 Off-delay timer   Figure 6.11 shows how a delay-on timer can be used to produce a *delay-off timer*. With such an arrangement, when there is a momentary input to In 1, both the output Out 1 and the timer are switched on. Because the input is latched by the Out 1 contacts, the output remains on. After the preset timer time delay, the timer contacts, which are normally closed, open and switch off the output. Thus the output starts as on and remains on until the time delay has elapsed.

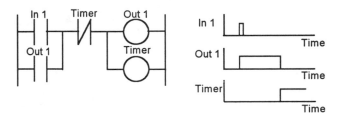

Figure 6.11 *Delay-off timer*

Some PLCs have, as well as on-delay timers, built-in off-delay timers and thus there is no need to use a on-delay timer to produce an off-delay timer. Figure 6.12 illustrates this for a Siemens PLC, giving the ladder diagram and the instruction list. Note that with this manufacturer, the timer is considered to be a delay item in a rung, rather than as a relay. In the rectangle symbol used for the timer, the 0 precedes the T and indicates that it is an on-delay timer.

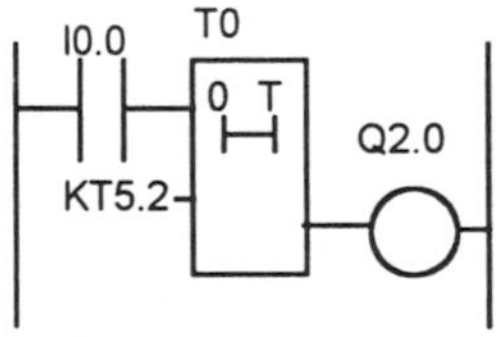

Figure 6.12 *Off-delay timer*

As an illustration of the use of an off-delay timer, consider the Allen Bradley program shown in Figure 6.13. TOF is used to indicate that it is an off-delay, rather than on-delay (TON) timer. The time base is set to 1:0 which is 1 s. The preset is 10 so the timer is preset to 10 s. In the first rung, the output of the timer is taken from the EN (for enable) contacts. This means that there is no time delay between an input to I:012/01 and the EN output. As a result the EN contacts in rung 2 close immediately there is an I:012/01 input. Thus there is an output from O:013/01 immediately the input I:012/01 occurs. The TT (for timer timing) contacts in rung 3 are energised just while the timer is running. Because the timer is an off-delay timer, the timer is turned on for 10 s before turning off. Thus the TT contacts will close when the set time of 10 s is running. Hence output O:012/02 is switched on for this time of 10 s. The DN (for done) contacts which are normally closed, open after the 10 s and so output O:013/03 comes on after 10 s. The DN contacts which are normally open, close after 10 s and so output O:013/04 goes off after 10 s.

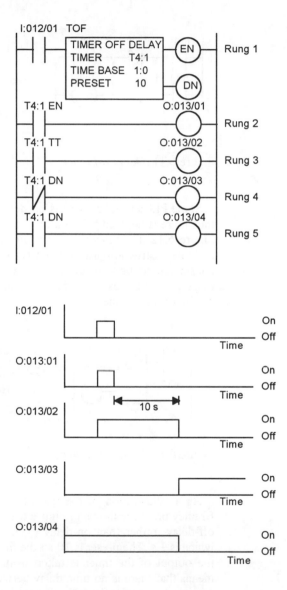

Figure 6.13 *Application of an off-delay timer*

6.4 One-shot timers

One-shot timers are used to produce a fixed duration output from some initiating input. Figure 6.14(a) shows a ladder diagram for a system that will give an output from Out 1 for a predetermined fixed length of time when there is an input to In 1, the timer being one involving a coil. There are two outputs for the input In 1. When there is an input to In 1, there is an output from Out 1 and the timer starts. When the predetermined time has elapsed, the timer contacts open. This switches off the output. Thus the output remains on for just the time specified by the timer.

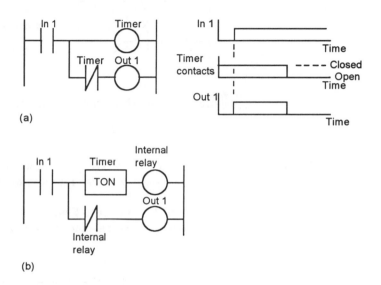

Figure 6.14 *One-shot on-timer*

Figure 6.14(b) shows an equivalent ladder diagram to Figure 6.13(a) but employing a timer which produces a delay in the time taken for a signal to reach the output.

In Figure 6.14, the one-shot timer has an output switched on by an input for a predetermined time, then switching off. Figure 6.15 shows another one-shot timer that switches an output on for a predetermined time after the input ceases. This uses a timer and two internal relays. When there is an input to In 1, the internal relay IR 1 is energised. The timer does not start at this point because the normally closed In 1 contacts are open. The closing of the IR 1 contacts means that the internal relay IR 2 is energised. There is, however, no output from Out 1 at this stage because, for the bottom rung, we have In 1 contacts open. When the input to In 1 ceases, both the internal relays remain energised and the timer is started. After the set time, the timer contacts, which are normally closed, open and switch off IR 2. This in turn switches off IR 1. It also, in the bottom rung, switches off the output Out 1. Thus the output is off for the duration of the input, then being switched on for a predetermined length of time.

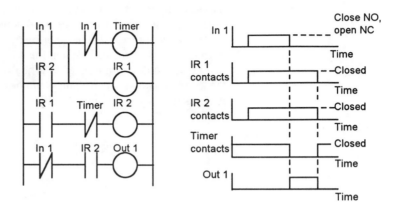

Figure 6.15 *One-shot timer on, when output ceases*

Problems

Questions 1 to 19 have four answer options A, B, C or D. Choose the correct answer from the answer options.

Problems 1 to 3 refer to Figure 6.16 which shows a ladder diagram with an on-delay timer, an input In 1 and an output Out 1.

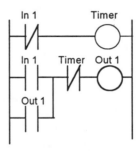

Figure 6.16 *Problems 1 to 3*

1 Decide whether each of these statements is True (T) or False (F).

When there is an input to In 1:
(i) The timer starts.
(ii) There is an output from Out 1.

A (i) T (ii) T
B (i) T (ii) F
C (i) F (ii) T
D (i) F (ii) F

2   Decide whether each of these statements is True (T) or False (F).

The timer starts when:
(i) There is an output.
(ii) The input ceases.

A (i) T (ii) T
B (i) T (ii) F
C (i) F (ii) T
D (i) F (ii) F

3   Decide whether each of these statements is True (T) or False (F).

When there is an input to In 1, the output is switched:
(i) On for the time for which the timer was preset.
(ii) Off for the time for which the timer was preset.

A (i) T (ii) T
B (i) T (ii) F
C (i) F (ii) T
D (i) F (ii) F

Problems 4 to 6 refer to Figure 6.17 which shows two alternative versions of a ladder diagram with two inputs In 1 and In 2, two outputs Out 1 and Out 2 and an on-delay timer.

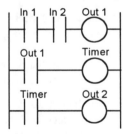 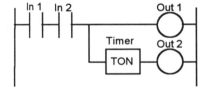

Figure 6.17 *Problems 4 to 6*

4   Decide whether each of these statements is True (T) or False (F).

When there is just an input to In 1:
(i) The timer starts.
(ii) There is an output from Out 2.

A (i) T (ii) T
B (i) T (ii) F
C (i) F (ii) T
D (i) F (ii) F

5 Decide whether each of these statements is True (T) or False (F).

When there is just an input to In 2:
(i) The timer starts.
(ii) There is an output from Out 2.

A (i) T (ii) T
B (i) T (ii) F
C (i) F (ii) T
D (i) F (ii) F

6 Decide whether each of these statements is True (T) or False (F).

When there is an input to In 1 and no input to In 2, there is an output from Out 2 which:
(i) Starts immediately.
(ii) Ceases after the timer preset time.

A (i) T (ii) T
B (i) T (ii) F
C (i) F (ii) T
D (i) F (ii) F

7 The program instruction list for a Mitsubishi PLC is: LD X400, OUT T450, K 6, LD T450, OUT Y430. An input to X400 gives:

A An output which is on for 6 s then off for 6 s.
B An output which lasts for 6 s.
C An output which starts after 6 s.
D An output which is off for 6 s, then on for 6 s.

8 The program instruction list for a Telemecanique PLC is: L I0,0, = T0, L T0, = O0,0. When there is an input to I0,0 there is:

A An output which is on for 6 s then off for 6 s.
B An output which lasts for 6 s.
C An output which starts after 6 s.
D An output which is off for 6 s, then on for 6 s.

Problems 9 and 10 refer to the program instruction list for a Mitsubishi PLC: LD X400, OR Y430, ANI T450, OUT Y430, LD X401, OR M100, AND Y430, OUT T450, K 10, OUT M100.

9 Decide whether each of these statements is True (T) or False (F).

When there is a momentary input to X400:
(i) The output from Y430 starts.
(ii) The output from Y430 ceases after 10 s.

A (i) T (ii) T
B (i) T (ii) F
C (i) F (ii) T
D (i) F (ii) F

10 Decide whether each of these statements is True (T) or False (F).

The output from Y430:
(i) Starts when there is a momentary input to X401.
(ii) Ceases 10 s after the input to X401.

A (i) T (ii) T
B (i) T (ii) F
C (i) F (ii) T
D (i) F (ii) F

Problems 11 and 12 refer to Figure 6.18 which shows a system with an input In 1, an on-delay timer and an output Out 1. The timer is set for a time of 6 s. The graph shows how the signal to the input varies with time.

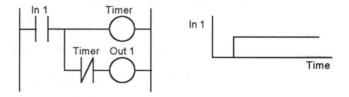

Figure 6.18 *Problems 11 and 12*

11 Decide whether each of these statements is True (T) or False (F).

The output from Out 1:
(i) Starts when the input starts.
(ii) Ceases 6 s after the start of the input.

A (i) T (ii) T
B (i) T (ii) F
C (i) F (ii) T
D (i) F (ii) F

12 Decide whether each of these statements is True (T) or False (F).

The timer contacts:
(i) Remain closed for 5 s after the start of the input.
(ii) Open 5 s after the input starts.

A (i) T (ii) T
B (i) T (ii) F
C (i) F (ii) T
D (i) F (ii) F

Questions 13 to 15 refer to Figure 6.19 which shows a ladder program for a Toshiba PLC involving internal relays, denoted by the letter R, and a TON timer with a preset of 20 s.

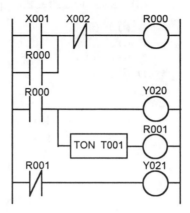

Figure 6.19 *Problems 13 to 15*

13 Decide whether each of these statements is True (T) or False (F).

The internal relay R000:
(i) Is used to latch the input X001.
(ii) Is used to start the timer T001.

A (i) T (ii) T
B (i) T (ii) F
C (i) F (ii) T
D (i) F (ii) F

14 Decide whether each of these statements is True (T) or False (F).

With no input to X002, the output Y020 is:
(i) Energised when there is an input to X001.
(ii) Ceases when there is no input to X001.

A (i) T (ii) T
B (i) T (ii) F
C (i) F (ii) T
D (i) F (ii) F

15 Decide whether each of these statements is True (T) or False (F).

With no input to X002:
(i) The output Y021 is switched on 20 s after the input X001.
(ii) The output Y020 is switched off 20 s after the input X001.

A (i) T (ii) T
B (i) T (ii) F
C (i) F (ii) T
D (i) F (ii) F

Questions 16 to 19 refer to Figure 6.20 which shows an Allen Bradley program and Figure 6.21 which shows a number of time charts for a particular signal input to I:012/01.

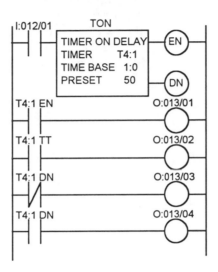

Figure 6.20 *Problems 16 to 19*

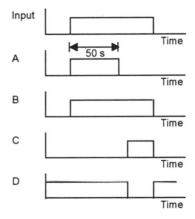

Figure 6.21 *Problems 16 to 19*

16 For the input shown in Figure 6.21, which is the output from O:013/01?

17 For the input shown in Figure 6.21, which is the output from O:013/02?

18 For the input shown in Figure 6.21, which is the output from O:013/03?

19 For the input shown in Figure 6.21, which is the output from O:013/04?

20 Devise ladder programs for systems that will carry out the following tasks:
(a) Switch on an output 5 s after receiving an input and keep it on for the duration of that input.
(b) Switch on an output for the duration of the input and then keep it on for a further 5 s.
(d) Switch on an output for 5 s after the start of an input signal.

7 Counters

A *counter* allows a number of occurrences of input signals to be counted. This might be in a situation where items pass along a conveyor belt and a specified number have to be diverted into a box (see Figure 1.1(b)). It might be counting the number of revolutions of a shaft, or perhaps the number of people passing through a door. Counters for such applications are provided as built-in elements in PLCs. This chapter describes how such counters can be programmed.

7.1 Forms of counter

A counter is set to some preset number value and, when this value of input pulses has been received, it will operate its contacts. Thus normally open contacts would be closed, normally closed contacts opened.

There are two types of counter, though PLCs may not include both types. These are down-counters and up-counters. *Down-counters* count down from the preset value to zero, i.e. events are subtracted from the set value. When the counter reaches the zero value, its contacts change state. Most PLCs offer down counting. *Up-counters* count from zero up to the preset value, i.e. events are added until the number reaches the preset value. When the counter reaches the set value, its contacts change state.

Different PLC manufacturers deal with counters in slightly different ways. Some count down (CTD), or up (CTU), and reset (RES) and treat the counter as though it is a relay coil and so a rung output. In this way, counters can be considered to consist of two basic elements: one relay coil to count input pulses and one to reset the counter, the associated contacts of the counter being used in other rungs. Figure 7.1(a) illustrates this. Mitsubishi is an example of this type of manufacturer. Others treat the counter as an intermediate block in a rung from which signals emanate when the count is attained. Figure 7.1(b) illustrates this. Siemens is an example of this type of manufacturer. CTD indicates the count down element, RST the reset element and CV the count value.

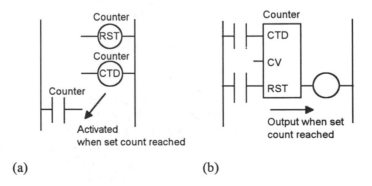

(a)   (b)

Figure 7.1 *Forms of representation of counters*

7.2 Programming

Figure 7.2 shows a basic counting circuit. When there is a pulse input to In 1, the counter is reset. When there is an input to In 2, the counter starts counting. If the counter is set for, say, 10 pulses, then when 10 pulse inputs have been received at In 2, the counter's contacts will close and there will be an output from Out 1. If at any time during the counting there is an input to In 1, the counter will be reset and start all over again and count for 10 pulses.

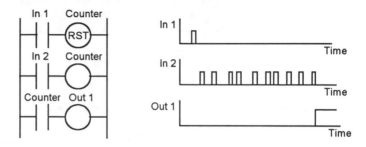

Figure 7.2 *Basic counter program*

Figure 7.3 shows how the above program, and its program instruction list, would appear with a Mitsubishi PLC. The reset and counting elements are combined in a single box spanning the two rungs. You can consider the rectangle to be enclosing the two counter O outputs in Figure 7.2. The count value is set by a K program instruction.

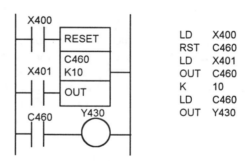

Figure 7.3 *Mitsubishi program*

Figure 7.4 shows the same program, and its program instruction list, with a Siemens PLC. With this ladder program, the counter is considered to be a delay element in the output line (as in Figure 7.1(b)). The counter is reset by an input to I0.2 and counts the pulses into input I0.1. The CU indicates that it is a count-up counter, a CD would indicate a count-down counter. The counter set value is indicated by the KC number.

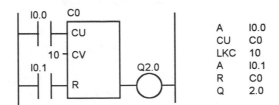

Figure 7.4 *Siemens program*

Figure 7.5 shows some more versions of the above ladder program for other manufacturers. Figure 7.5(a) is for Toshiba and (b) for Allen Bradley. Figure 7.6 shows how the outputs DN and CU are controlled by the count for the Allen Bradley configuration.

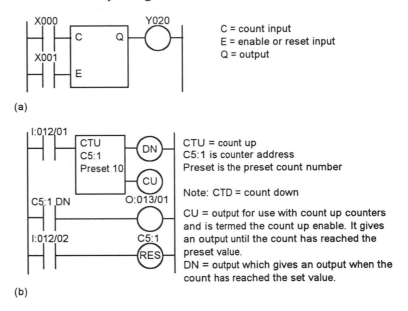

Figure 7.5 *(a) Toshiba, (b) Allen Bradley programs*

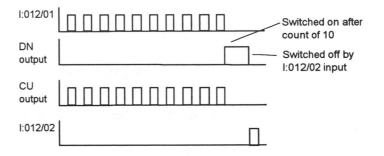

Figure 7.6 *Count chart for Allen Bradley program*

7.2.1 Counter application

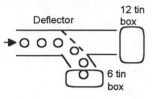

Figure 7.7 *Counting task*

As an illustration of the use that can be made of a counter, consider the problem of the control of a machine which is required to direct 6 tins along one path for packaging in a box and then 12 tins along another path for packaging in another box (Figure 7.7). A deflector plate might be controlled by a photocell sensor which gives an output every time a tin passes it. Thus the number of pulses from the sensor has to be counted and used to control the deflector. Figure 7.8 shows the ladder program that could be used. Mitsubishi notation has been used.

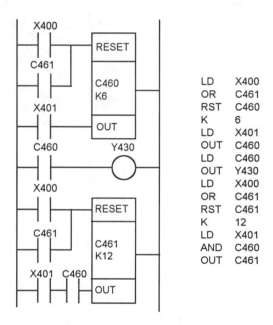

Figure 7.8 *Ladder program for Figure 7.7 task*

When there is a pulse input to X400, both the counters are reset. The input to X400 could be the push button switch used to start the conveyor moving. The input which is counted is X401. This might be an input from a photocell sensor which detects the presence of tins passing along the conveyor. C460 starts counting after X400 is momentarily closed. When C460 has counted six items, it closes its contacts and so gives an output at Y430. This might be a solenoid which is used to activate a deflector to deflect items into one box or another. Thus the deflector might be in such a position that the first six tins passing along the conveyor are deflected into the 6-pack box, then the deflector plate is moved to allow tins to pass to the 12-pack box. When C460 stops counting it closes its contacts and so allows C461 to start counting. C461 counts for 12 pulses to X401 and then closes its contacts. This results in both counters being reset and the entire process can repeat itself.

7.3 Up and down counting

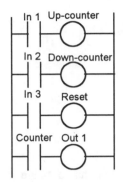

Figure 7.9 *Using up- and down-counters*

It is possible to program up- and down-counters together. Consider the task of counting products as they enter a conveyor line and as they leave it, or perhaps cars as they enter a multi-storage parking lot and as they leave it. An output is to be triggered if the number of items/cars entering is some number greater than the number leaving, i.e. the number in the parking lot has reached a 'saturation' value. The output might be to illuminate a 'No empty spaces' sign. Suppose we use the up-counter for items entering and the count down for items leaving. Figure 7.9 shows the basic form a ladder program for such an application can take. When an item enters it gives a pulse on input In 1. This increases the count by one. Thus each item entering increases the accumulated count by 1. When an item leaves it gives an input to In 2. This reduces the number 1. Thus each item leaving reduces the accumulated count by 1. When the accumulated value reaches the preset value, the output Out 1 is switched on.

Figure 7.10 shows how the above system might appear for a Siemens PLC and the associated program instruction list. CU is the count up input and CD the count down. R is the reset. The set accumulator value is loaded via F0.0, this being an internal relay.

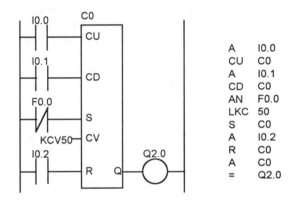

Figure 7.10 *Up and down counting with a Siemens PLC*

7.4 Sequencers

The *drum sequencer* is a form of counter that is used for sequential control. It replaces the mechanical drum sequencer that was used to control machines that have a stepped sequence of repeatable operations. One form of the mechanical drum sequencer consisted of a drum from which a number of pegs protruded (Figure 7.11). When the cylinder rotated, contacts aligned with the pegs were closed when the peg impacted on them and opened when the peg had passed. Thus for the arrangement shown in Figure 7.11, as the drum rotates, in the first step the peg for output 1 is activated, in step 2 the peg for the third output, in step 3 the peg for the second output, and so on. Different outputs could be controlled by pegs located at different distances along the drum. Another form consisted of a series of cams on the same shaft, the profile of the cam being used to switch contacts on and off.

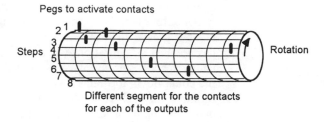

Figure 7.11 *Drum sequencer*

The PLC sequencer consists of a master counter that has a range of presets counts corresponding to the different steps and so, as it progresses through the count, when each preset count is reached can be used to control outputs. Each step in the count sequence relates to a certain output or group of outputs. The outputs are internal relays, these in turn being used to control the external output devices.

Suppose we want output 1 to be switched on 5 s after the start and remain on until the time reaches 10 s, output 2 to be switched on at 10 s and remain on until 20 s, output 3 to be switched on at 15 s and remain on until 25 s, etc. We can represent the above requirements by a time sequence diagram, Figure 7.12 showing the required time sequence.

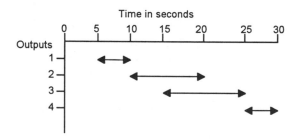

Figure 7.12 *Timing diagram*

We can transform the timing diagram into a drum sequence requirement. Taking each drum sequence step to take 5 s gives the requirement diagram shown in Figure 7.13. Thus at step 1 we require output 1 to be switched on and to remain on until step 2. At step 2 we require output 2 to be switched on and remain on until step 4. At step 3 we require output 3 to be switched on and remain on until step 5. At step 5 we require output 4 to be switched on and remain on until step 6.

Step	Time (s)	Output			
		1	2	3	4
0	0	0	0	0	0
1	5	1	0	0	0
2	10	0	1	0	0
3	15	0	1	1	0
4	20	0	0	1	0
5	25	0	0	1	1
6	30	0	0	0	0

Figure 7.13 *Sequence requirements*

The way in which drum sequencers are implemented depends on the manufacturer. One way of implementing the above sequence is to use a timer which is programmed to self-reset every 5 s after being activated by an input being closed. Figure 7.14 shows the form such a sequencer implementation might take.

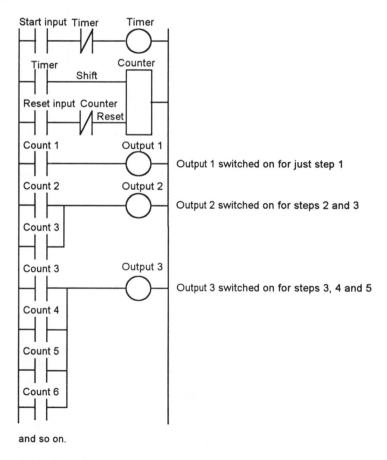

and so on.

Figure 7.14 *Use of a sequencer*

The timer contacts are used to switch on the sequencer counter which activates associated contacts, contacts 1 after one step, contacts 2 after two steps, contacts 3 after three steps, etc. Thus after step 1 the step 1 contacts are closed and the first input is activated and remains on for the duration of that step. After step 2 the second output is switched on and remains on for that step and the third step because the step 2 and step 3 contacts are in an OR arrangement. After step 3 the third output is switched on and remains on for steps 3, 4 and 5 because the 3, 4 and 5 contacts are in an OR arrangement.

With a PLC, such as the Mitsubishi F2, counters are designated as sequence counters by switching on the special relay M577. Thus counter C666 is designated as the sequence counter and counter C667 as the timer to set up the process transfer times for the internal relays M300-377. Figure 7.15 shows such a ladder program.

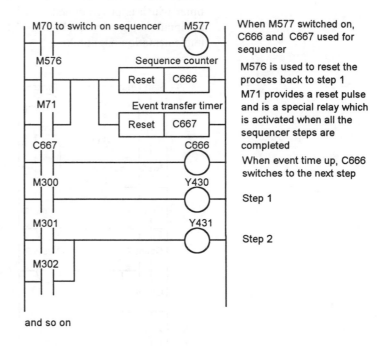

Figure 7.15 *Sequencer with a Mitsubishi F2*

With a PLC, such as a Toshiba, the sequencer is set up by switching on the Step Sequence Initialize (STIZ) function block R500 (Figure 7.16). This sets up the program for step 1 and R501. This relay then switches on output Y020. The next step is the switching on of R502. This switches on the output Y021 and also a delay-on timer so that R503 is not switched on until the timer has timed out. Then R503 switches on Y022 and also the next step in the sequence.

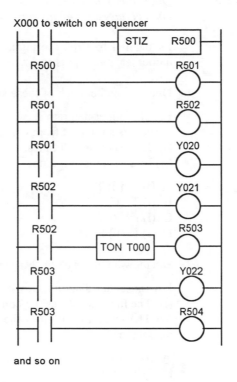

Figure 7.16 *Sequencer with a Toshiba PLC*

With the Allen Bradley form of PLC the sequencer is programmed by using a sequence of binary words in the form of the outputs required, e.g. those listed in Figure 7.13. Thus we would have the binary word sequence:

```
Input 4
  Input 3
    Input 2
     Input 1
0000
0001
0010
0110
0100
0100
0000
```

The binary words are inserted into the program using the programming device.

Problems

Questions 1 to 16 have four answer options A, B, C or D. Choose the correct answer from the answer options.

Problems 1 to 3 refer to Figure 7.17 which shows a ladder diagram with a counter, two inputs In 1 and In 2 and an output Out 1.

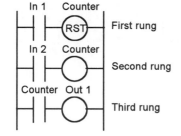

Figure 7.17 *Problems 1 to 3*

1 Decide whether each of these statements is True (T) or False (F).

For the ladder diagram shown in Figure 7.17, when the counter is set to 5, there is an output from Out 1 every time:
(i) In 1 has closed 5 times.
(ii) In 2 has closed 5 times.

A (i) T (ii) T
B (i) T (ii) F
C (i) F (ii) T
D (i) F (ii) F

2 Decide whether each of these statements is True (T) or False (F).

For the ladder diagram shown in Figure 7.17:
(i) The first rung gives the condition required to reset the counter.
(ii) The second rung gives the condition required to generate pulses to be counted.

A (i) T (ii) T
B (i) T (ii) F
C (i) F (ii) T
D (i) F (ii) F

3 Decide whether each of these statements is True (T) or False (F).

When there is an input to In 1:
(i) The counter contacts in the third rung close.
(ii) The counter is ready to start counting the pulses from In 2.

A (i) T (ii) T
B (i) T (ii) F
C (i) F (ii) T
D (i) F (ii) F

Problems 4 and 5 refer to the following program instruction list involving a down-counter:

LD	X400
RST	C460
LD	X401
OUT	C460
K	5
LD	460
OUT	Y430

4   Decide whether each of these statements is True (T) or False (F).

Every time there is an input to X401:
(i) The count accumulated by the counter decreases by 1.
(ii) The output is switched on.

A (i) T (ii) T
B (i) T (ii) F
C (i) F (ii) T
D (i) F (ii) F

5   Decide whether each of these statements is True (T) or False (F).

When there is an input to X400, the counter:
(i) Resets to a value of 5.
(ii) Starts counting from 0.

A (i) T (ii) T
B (i) T (ii) F
C (i) F (ii) T
D (i) F (ii) F

Problems 6 and 7 refer to the following program instruction list involving a counter C0:

```
A      I0.0
CD     C0
LKC    5
A      I0.1
R      C0
Q      2.00
```

6   Decide whether each of these statements is True (T) or False (F).

Every time there is an input to I0.0:
(i) The count accumulated by the counter decreases by 1.
(ii) The output is switched on.

A (i) T (ii) T
B (i) T (ii) F
C (i) F (ii) T
D (i) F (ii) F

7   Decide whether each of these statements is True (T) or False (F).

When there is an input to I0.1, the counter:
(i) Resets to a value of 5.
(ii) Starts counting from 0.

A (i) T (ii) T
B (i) T (ii) F
C (i) F (ii) T
D (i) F (ii) F

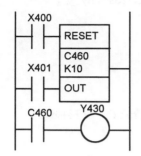

Figure 7.18 *Problems 8 and 9*

Problems 8 and 9 refer to Figure 7.18 which shows a down-counter C460 controlled by two inputs X400 and X401, there being an output from Y430.

8 Decide whether each of these statements is True (T) or False (F).

When there is an input to X400, the counter:
(i) Resets to a value of 0.
(ii) Starts counting.

A (i) T (ii) T
B (i) T (ii) F
C (i) F (ii) T
D (i) F (ii) F

9 Decide whether each of these statements is True (T) or False (F).

Every time there is an input to X401, the counter:
(i) Gives an output from Y430.
(ii) Reduces the accumulated count by 1.
A (i) T (ii) T
B (i) T (ii) F
C (i) F (ii) T
D (i) F (ii) F

Problems 10 to 12 refer to Figure 7.19 which shows a ladder diagram involving a counter C460, inputs X400 and X401, internal relays M100 and M101, and an output Y430.

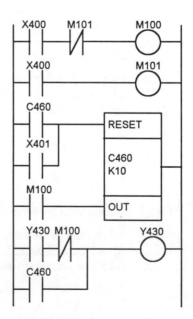

Figure 7.19 *Problems 10 to 12*

10 Decide whether each of these statements is True (T) or False (F).

For the output Y430:
(i) It switches on the tenth pulse to X400.
(ii) It switches off at the start of the eleventh pulse to X400.

A (i) T (ii) T
B (i) T (ii) F
C (i) F (ii) T
D (i) F (ii) F

11 Decide whether each of these statements is True (T) or False (F).

When there is an input to X400:
(i) The internal relay M100 is energised.
(ii) The internal relay M101 is energised.

A (i) T (ii) T
B (i) T (ii) F
C (i) F (ii) T
D (i) F (ii) F

12 Decide whether each of these statements is True (T) or False (F).

There is an output from Y430 as long as:
(i) The C460 contacts are closed.
(ii) Y430 gives an output and M100 is energised.

A (i) T (ii) T
B (i) T (ii) F
C (i) F (ii) T
D (i) F (ii) F

13 Decide whether each of these statements is True (T) or False (F).

Figure 7.20 shows a counter program in Siemens format. After 10 inputs to I0.0:
(i) The lamp comes on.
(ii) The motor starts.

A (i) T (ii) T
B (i) T (ii) F
C (i) F (ii) T
D (i) F (ii) F

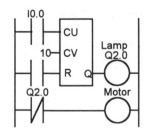

Figure 7.20 *Problem 13*

Questions 14 and 15 refer to Figure 7.21 which shows a Siemens program involving an up and down counter.

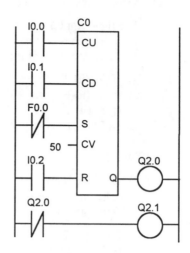

Figure 7.21 *Problems 14 and 15*

14 Decide whether each of these statements is True (T) or False (F).

When the count is less that 50:
(i) There is an output from Q2.0.
(ii) There is an output from Q2.1.

A (i) T (ii) T
B (i) T (ii) F
C (i) F (ii) T
D (i) F (ii) F

15 Decide whether each of these statements is True (T) or False (F).

When the count reaches 50:
(i) There is an output from Q2.0.
(ii) There is an output from Q2.1.

A (i) T (ii) T
B (i) T (ii) F
C (i) F (ii) T
D (i) F (ii) F

16 For the Allen Bradley program shown in Figure 7.22, the counter is reset when:

A The count reaches 5.
B The count passes 5.
C There is an input to I:012/01.
D There is an input to I:012/02.

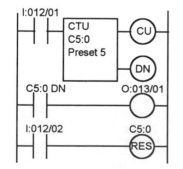

Figure 7.22 *Problem 16*

17 Devise ladder programs for systems that will carry out the following tasks:

(a) Give an output after a photocell sensor has given 10 pulse input signals as a result of detecting 10 objects passing in front of it.

(b) Give an output when the number of people in a store reaches 100, there continually being people entering and leaving the store.

8 Shift registers

The term *register* is used for an electronic device in which data can be stored. An internal relay, see Chapter 5, is such a device. The *shift register* is a number of internal relays grouped together which allow stored bits to be shifted from one relay to another. This chapter is about shift registers and how they can be used where a sequence of operations is required or to keep track of particular items in a production system.

8.1 Shift registers

A register is a number of internal relays grouped together, normally 8, 16 or 32. Each internal relay is either effectively open or closed, these states being designated as 0 and 1. The term *bit* is used for each such binary digit. Therefore, if we have eight internal relays in the register we can store eight 0/1 states. Thus we might have:

Internal relays

1	2	3	4	5	6	7	8

and each relay might store an on–off signal such that the state of the register at some instant is:

1	0	1	1	0	0	1	0

i.e. relay 1 is on, relay 2 is off, relay 3 is on, relay 4 is on, relay 5 is off, etc. Such an arrangement is termed an 8-bit register. Registers can be used for storing data that originate from input sources other than just simple, single on–off devices such as switches.

With the *shift register* it is possible to shift stored bits. Shift registers require three inputs, one to load data into the first location of the register, one as the command to shift data along by one location and one to reset or clear the register of data. To illustrate this, consider the following situation where we start with an 8-bit register in the following state:

1	0	1	1	0	0	1	0

Suppose we now receive the input signal 0. This is an input signal to the first internal relay.

Input 0

→ | 1 | 0 | 1 | 1 | 0 | 0 | 1 | 0 |

If we also receive the shift signal, then the input signal enters the first location in the register and all the bits shift along one location. The last bit overflows and is lost.

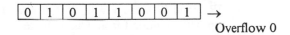

$$0 \mid 1 \mid 0 \mid 1 \mid 1 \mid 0 \mid 0 \mid 1 \ \rightarrow$$

Overflow 0

Thus a set of internal relays that were initially on, off, on, on, off, off, on, off are now off, on, off, on, on, off, off, on.

The grouping together of internal relays to form a shift register is done automatically by a PLC when the shift register function is selected. With the Mitsubishi PLC, this is done by using the programming code SFT (shift) against the internal relay number that is to be the first in the register array. This then causes a block of relays, starting from that initial number, to be reserved for the shift register.

8.2 Ladder programs

Consider a 4-bit shift register and how it can be represented in a ladder program (Figure 8.1). The input In 3 is used to reset the shift register, i.e. put all the values at 0. The input In 1 is used to input to the first internal relay in the register. The input In 2 is used to shift the states of the internal relays along by one. Each of the internal relays in the register, i.e. IR 1, IR 2, IR 3 and IR 4, is connected to an output, these being Out 1, Out 2, Out 3 and Out 4.

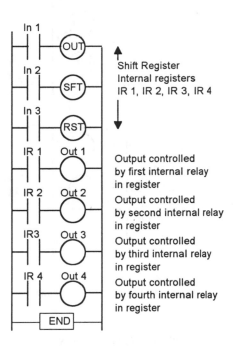

Figure 8.1 *The shift register*

Suppose we start by supplying a momentary input to In 3. All the internal relays are then set to 0 and so the states of the four internal relays IR 1, IR 2, IR 3 and IR 4 are 0, 0, 0, 0. When In 1 is momentarily closed there is a 1 input into the first relay. Thus the states of the internal relays IR 1, IR 2,

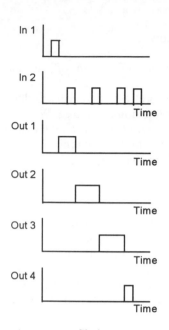

Figure 8.2 *Shift register*

IR 3 and IR 4 are now 1, 0, 0, 0. The IR 1 contacts close and we thus end up with an output from Out 1. If we now supply a momentary input to In 2, the 1 is shifted from the first relay to the second. The states of the internal relays are now 0, 1, 0, 0. We now have no input from Out 1 but an output from Out 2. If we supply another momentary input to In 2, we shift the states of the relays along by one location to give 0, 0, 1, 0. Outputs 1 and 2 are now off but Out 3 is on. If we supply another momentary input to In 2 we again shift the states of the relays along by one and have 0, 0, 0, 1. Thus now, outputs 1, 2 and 3 are off and output 4 has been switched on. When another momentary input is applied to In 2, we shift the states of the relays along by one and have 0, 0, 0, 0 with the 1 overflowing and being lost. All the outputs are then off. Thus the effect of the sequence of inputs to In 2 has been to give a sequence of outputs Out 1, followed by Out 2, followed by Out 3, followed by Out 4. Figure 8.2 shows the sequence of signals.

Figure 8.3 shows the Mitsubishi version of the above ladder program and the associated instruction list.

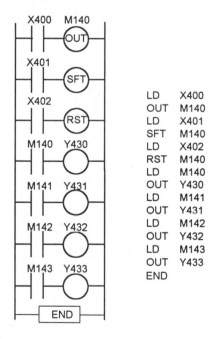

Figure 8.3 *Mitsubishi program*

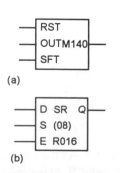

Figure 8.4 *(a) Mitsubishi, (b) Toshiba*

Figure 8.4(a) shows how, instead of the three separate outputs for reset, output and shift shown above, the Mitsubishi shift register might appear in a program and Figure 8.4(b) shows the comparable element for Toshiba. With the Mitsubishi shift register, the M140 is the address of the first relay in the register. With the Toshiba R016 is the address of the first relay in the register. The (08) indicates that there are eight such relays. D is used for the data input, S for shift input, E for enable or reset input and Q for output. As an illustration, Figure 8.5 shows a shift register ladder program for a Toshiba PLC.

8.2.1 Keeping track of items

The above indicates how a shift register can be used for sequencing. Another application is to keep track of items. For example, a sensor might be used to detect faulty items moving along a conveyor and keep track of it so that when it reaches the appropriate point a reject mechanism is activated to remove it from the conveyor. Figure 8.6 illustrates this arrangement and the type of ladder program that might be used.

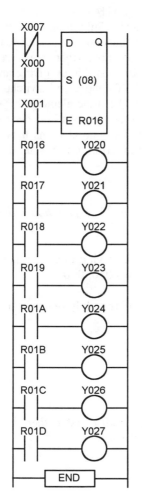

Figure 8.5 *Shift register*

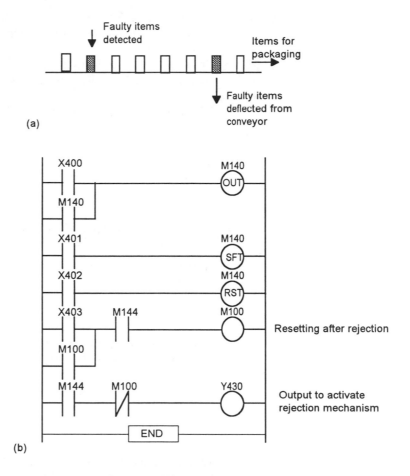

Figure 8.6 *Keeping track of faulty items*

Each time a faulty item is detected, a pulse signal occurs at input X400. This enters a 1 into the shift register at internal relay M140. When items move, whether faulty or not, there is a pulse input at X401. This shifts the 1 along the register. When the 1 reaches internal relay M144, it activates the output Y430 and the rejection mechanism removes the faulty item from the conveyor. When an item is removed it is sensed and an input to X403 occurs. This is used to reset the mechanism so that no further items are rejected until the rejection signal reaches M144. It does this by giving an output to internal relay M100 which latches the X403 input and switches the rejection output Y430 off. This represents just the basic elements of a system. A practical system would include further internal relays in order to make certain that the rejection mechanism is off when good items move along the conveyor belt and also to disable the input from X400 when the shifting is occurring.

Problems

Questions 1 to 9 have four answer options A, B, C or D. Choose the correct answer from the answer options.

Problems 1 to 5 concern a 4-bit shift register, involving internal relays IR 1, IR 2, IR 3, and IR 4, which has been reset to 0, 0, 0, 0.

1 When there is a pulse 1 input to the OUT of the shift register, the internal relays in the shift register show:

A 0001
B 0010
C 0100
D 1000

2 Following a pulse input of 1 to the OUT of the shift register, there is a pulse input to SHIFT. The internal relays then show:

A 0001
B 0010
C 0100
D 1000

3 With a continuous input of 1 to the OUT of the shift register, there is a pulse input to SHIFT. The internal relays then show:

A 0011
B 0110
C 1100
D 0010

4 With a continuous input of 1 to the OUT of the shift register, there are two pulse inputs to SHIFT. The internal relays then show:

A 0001
B 0010
C 1100
D 1110

5 With a pulse input of 1 to the OUT of the shift register, there is a pulse input to SHIFT, followed by a pulse input to RESET. The internal relays then show:

A 0000
B 0010
C 0100
D 1000

Problems 6 to 9 concern Figure 8.7 which shows a 4-bit shift register with internal relays IR 1, IR 2, IR 3 and IR 4, with three inputs In 1, In 2 and In 3, and four outputs Out 1, Out 2, Out 3 and Out 4.

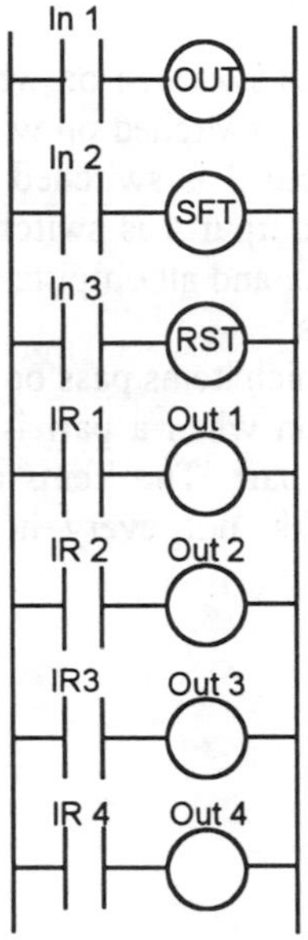

6 Decide whether each of these statements is True (T) or False (F).

When there is a pulse input to In 1:
(i) The output Out 1 is energised.
(ii) The contacts of the internal relay IR 1 close.

A (i) T (ii) T
B (i) T (ii) F
C (i) F (ii) T
D (i) F (ii) F

7 Decide whether each of these statements is True (T) or False (F).

When there is a pulse input to In 1 followed by a pulse input to SFT:
(i) The output Out 1 is energised.
(ii) The output Out 2 is energised.

A (i) T (ii) T
B (i) T (ii) F
C (i) F (ii) T
D (i) F (ii) F

Figure 8.7 *Problems 6 to 8*

8 Decide whether each of these statements is True (T) or False (F).

To obtain the outputs Out 1, Out 2, Out 3 and Out switching on in sequence and remaining on, we can have for inputs:
(i) A pulse input to In 1 followed by three pulse inputs to SFT.
(ii) A continuous input to In 1 followed by three pulse inputs to SFT.

A (i) T (ii) T
B (i) T (ii) F
C (i) F (ii) T
D (i) F (ii) F

9 To obtain the sequence:
Initially: Out 1 off, Out 2 off, Out 3 off, Out 4 off
Next: Out 1 on, Out 2 off, Out 3 off, Out 4 off
Next: Out 1 off, Out 2 on, Out 3 off, Out 4 off
Next: Out 1 on, Out 2 off, Out 3 on, Out 4 off

the inputs required are:

A   Pulse input to In 1 followed by pulse input to In 2.
B   Pulse input to In 1 followed by two pulses to In 2.
C   Pulse input to In 1 followed by pulse input to In 2, followed by pulse input to In 1.
D   Pulse input to In 1 followed by pulse input to In 2, followed by pulse inputs to In 1 and In 2.

10 Devise ladder programs for systems that will carry out the following tasks:

(a) A sequence of four outputs such that output 1 is switched on when the first event is detected and remains on, output 2 is switched on when the second event is detected and remains on, output 3 is switched on when the third event is detected and remains on, output 4 is switched on when the fourth event is detected and remains on, and all outputs are switched off when one particular input signal occurs.

(b) Control of a paint sprayer in a booth through which items pass on an overhead conveyor so that the paint is switched on when a part is in front of the paint gun and off when there is no part. The items are suspended from the overhead conveyor by hooks, not every hook having an item suspended from it.

9 Data handling

Timers, counters and individual internal relays are all concerned with the handling of individual bits, i.e. single on–off signals. Shift registers involve a number of bits with a group of internal relays being linked (see Chapter 8). The block of data in the register is manipulated. This chapter is about PLC operations involving blocks of data representing a value, such blocks being termed *words*. A block of data is needed if we are to represent numbers rather than just a single on–off input. Data handling consists of operations involving moving or transferring numeric information stored in one memory word location to another word in a different location, comparing data values and carrying out simple arithmetic operations. For example, there might be the need to compare a numeric value with a set value and initiate action if the actual value is less than the set value. This chapter is an introductory discussion of such operations.

9.1 Registers and bits

A register is where data can be stored (see section 8.1 for an initial discussion of registers). In a PLC there are a number of such registers. Each data register can store a *binary word* of usually 8 or 16 bits. The number of bits determines the size of the number that can be stored. The *binary system* uses only two symbols, 0 and 1. Thus we might have the 4-bit number 1111. The *least significant bit* (LSB) is the one at the right and a 1 in that position represents 2^0. A 1 in the next position along represents 2^1. In the next position 2^2 and in the fourth position from the right 2^3, this being the *most significant bit* (MSB) for this number. Thus 1111 is:

$$\begin{array}{cccc} \text{MSB} & & \text{LSB} & \\ 1 & 1 & 1 & 1 \\ 2^3 & 2^2 & 2^1 & 2^0 \end{array}$$

and so is the denary number, i.e. the familiar number system based on 10s, of $2^0 + 2^1 + 2^2 + 2^3 = 1 + 2 + 4 + 8 = 15$. Thus a 4-bit register can store a positive number between 0 and $2^0 + 2^1 + 2^2 + 2^3$ or $2^4 - 1 = 15$. An 8-bit register can store a positive number between 0 and $2^0 + 2^1 + 2^2 + 2^3 + 2^4 + 2^5 + 2^6 + 2^7$ or $2^8 - 1$, i.e. 255. A 16-bit register can store a positive number between 0 and $2^{16} - 1$, i.e. 65 535.

Thus a 16-bit word can be used for positive numbers in the range 0 to +65 535. If negative numbers are required, the most significant bit is used to represent the sign, a 1 representing a negative number and a 0 a positive number and the format used for the negative numbers is *two's complement*. Two's complement is a way of writing negative numbers so that when we add, say, the signed equivalent of +5 and −5 we obtain 0. Thus in this format, 1011 represents the negative number −5 and 0101 the positive

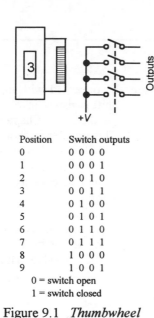

Position	Switch outputs
0	0 0 0 0
1	0 0 0 1
2	0 0 1 0
3	0 0 1 1
4	0 1 0 0
5	0 1 0 1
6	0 1 1 0
7	0 1 1 1
8	1 0 0 0
9	1 0 0 1

0 = switch open
1 = switch closed

Figure 9.1   *Thumbwheel switch*

number +5; 1011 + 0101 = (1)0000 with the (1) for the 4-bit number being lost. See the Appendix for further discussion.

The *binary coded decimal* (BCD) format is often used with PLCs when they are connected to devices such as digital displays. With the natural binary number there is no simple link between the separate symbols of a denary number and the equivalent binary number. You have to work out the arithmetic to establish one number from the other. With the binary coded decimal system, each denary digit is represented, in turn, by a 4-bit binary number (four is the smallest number of binary bits that gives a denary number greater than 10, i.e. $2^n > 10$). To illustrate this, consider the denary number 123. The 3 is represented by the 4-bit binary number 0011, the 2 by the 4-bit number 0010 and the 1 by 0001. Thus the binary coded decimal number of 123 is 0001 0010 0011. Binary coded decimal is a convenient system for use with external devices which are arranged in denary format, e.g. decade switches (thumbwheel switches) and digital displays. Then four binary bits can be used for each denary digit. PLCs therefore often have inputs or outputs which can be programmed to convert binary coded decimal from external input devices to the binary format needed for inside the PLC and from the binary format used internally in the PLC to binary coded decimal for external output devices (see Section 9.3).

The thumbwheel switch is widely used as a means of inputting BCD data manually into a PLC. It has four contacts which can be opened or closed to give the four binary bits to represent a denary number (Figure 9.1). The contacts are opened or closed by rotating a wheel using one's thumb. By using a number of such switches, data can be inputted in BCD format.

9.2 Data handling

For data handling the typical instruction will contain the data-handling instruction, the source (S) address from where the data is to be obtained and the destination (D) address to where it to be moved. Figure 9.2 shows a common format.

Figure 9.2 *Data-handling instruction*

The following are examples of data-handling instructions to be found with PLCs.

9.2.1 Data movement

The instruction commonly used to move data is MOV. This copies a value from one address to another. Figure 9.3 illustrates a common practice of using one rung of a ladder program for each move operation, showing the form used by two manufactures, Mitsubishi and Allen Bradley. For the rung shown, when there is an input to ‖ in the rung, the move occurs from the designated source address to the designated destination address. Another approach that is used by some manufactures, e.g. Siemens, is to regard data movement as two separate instructions, loading data from the source into an accumulator and then transferring the data from the accumulator to the destination.

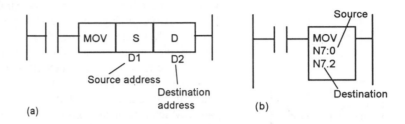

(a)   (b)

Figure 9.3 *Data movement: (a) Mitsubishi, (b) Allen Bradley*

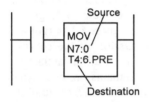

Figure 9.4 *Moving number to timer preset*

Data transfers might be to move a preset value to a timer or counter, or a time or counter value to some register for storage, or data from an input to a register or a register to output. Figure 9.4 shows the rung, in the Allen Bradley format, that might be used to transfer a number held at address N7:0 to the preset of timer T4:6 when the input conditions for that rung are met. A data transfer from the accumulated value in a counter to a register would have a source address of the form C5:18.ACC and a destination address of the form N7:0. A data transfer from an input to a register might have a source address of the form I:012 and a destination address of the form N7:0. A data transfer from a register to an output might have a source address of the form N7:0 and a destination address of the form O:030.

9.2.2 Data comparison

The data comparison instruction gets the PLC to compare two data values. Thus it might be to compare a digital value read from some input device with a second value contained in a register. For example, we might want some action to be initiated when the input from a temperature sensor gives a digital value which is less than a set value stored in a data register in the PLC. PLCs generally can make comparisons for *less than* (< or LES), *equal to* (= or EQU), *less than or equal to* (≤ or <= or LEQ), *greater than* (> or GRT), *greater than or equal to* (≥ or >= or GEQ) and *not equal to* (≠ or <> or NEQ). The brackets alongside each of the terms indicates common abbreviations used in programming.

For data comparison the typical instruction will contain the data-transfer instruction to compare data, the source (S) address from where the data is to be obtained for the comparison and the destination (D) address of the data against which it is to be compared. The instructions commonly used for the comparison are the terms indicated in the above brackets. When

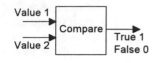

value 1 is compared with value 2 and the result agrees with the form of comparison used then the outcome is a true 1 output; if the comparison is not true then the outcome is false and so a 0 output (Figure 9.5). For example, using the less than form of comparison, if value 1 is less than value 2 the outcome is true and so 1. If it is not less than value 1, the outcome of the comparison is false and so the output is 0.

Figure 9.5 *Comparison*

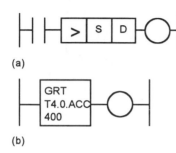

(a)

(b)

Figure 9.6 *Greater than comparison*

Figure 9.6 shows the type of formats used by two manufacturers using the greater than form of comparison. Similar forms apply to the other forms of comparison. In Figure 9.6(a) the format is that used by Mitsubishi, S indicating the source of the data value for the comparison and D the destination or value against which the comparison is to be made. Thus if the source value is greater than the destination value, the output is 1. In Figure 9.6(b) the Allen Bradley format has been used. Here the source of the data being compared is given as the accumulated value in timer 4.0 and the data against which it is being compared is the number 400.

As an illustration of the use of such a comparison, consider the task of sounding an alarm if a sensor indicates that a temperature has risen above some value, say 100°C. The alarm is to remain sounding until the temperature falls below 90°C. Figure 9.7 shows the ladder diagram that might be used. When the temperature rises to become equal to or greater than 100°C, then the greater than comparison element gives a 1 output and so sets an internal relay (see section 5.5 for a discussion of the set/reset form of internal relay). There is then an output. This output latches the greater than comparison element and so the output remains on, even when the temperature falls below 100°C. The output is not switched off until the less than 90°C gives an output and resets the internal relay.

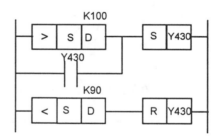

Figure 9.7 *Alarm program*

Another example of the use of comparison is when, say, four outputs need to be started in sequence, i.e. output 1 starts when the initial switch is closed, followed some time later by output 2, some time later by output 3 and some time later by output 4. While this could be done using three timers, another possibility is to use one timer with greater than or equal elements. Figure 9.8 shows a possible ladder diagram.

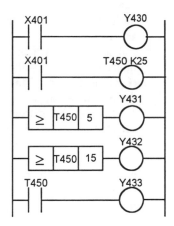

Figure 9.8 *Sequential switching on*

When the X400 contacts close the output Y430 starts. Also the timer is started. When the timer accumulated value reaches 5 s then the greater than or equal to element switches on Y431. When the timer accumulated value reaches 15 s then the greater than or equal to element switches on Y432. When the timer reaches 25 s then its contacts switch on Y433.

9.3 Arithmetic instructions

Most PLCs provide BCD-to-binary and binary-to-BCD conversions for use when the input might be for a thumbwheel switch or the output to a decimal display. Figure 9.9 shows the data handling instructions for use in such situations.

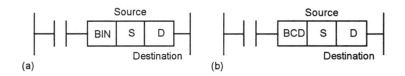

Figure 9.9 *Conversion: (a) BCD-to-binary, (b) binary-to-BCD*

9.3.1 Arithmetic operations

Some PLCs are equipped to carry out just the arithmetic operations of addition and subtraction, others the four basic arithmetic operations of addition, subtraction, multiplication and division, while others can carry out these and various other functions such as the exponential. Addition and subtraction operations are used to alter the value of data held in data registers. For example, this might be to adjust a sensor input reading or perhaps obtain a value by subtracting two sensor values or alter the preset values used by timers and counters. Multiplication might be used to multiply some input before perhaps adding to or subtracting it from another.

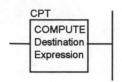

Figure 9.10 *Allen Bradley format*

The way in which PLCs have to be programmed to carry out such operations varies. Allen Bradley in some of their PLCs use a compute (CPT) instruction. This is an output instruction that performs the operations defined and then writes the results to a specified destination address. Figure 9.10 shows the instruction format. When the compute instruction is on the programming screen, the destination has to be first entered, then the expression. Thus we might have a destination of T4:1.ACC and an expression (N7:1 + N10:1)*3.5. Note that the symbol * is used for multiplication. The expression states that the value in N7:1 is to be added to the value in N10:1. This sum is then to be multiplied by 3.5. The result is then to be sent to the destination which is the accumulated value in timer 4.1.

9.4 Continuous control

Continuous control of some variable, e.g. the control of the temperature in a room, can be achieved by comparing the actual value for the variable with the desired set value and then giving an output, such as switching on a heater, to reduce the difference. Figure 9.11 illustrates this by means of a block diagram. The actual value of the variable is compared with the set value and a signal obtained representing the difference or error. A controller then takes this difference signal and gives an output to an actuator to give a response to correct the discrepancy. Such a system is said to be *closed-loop control*.

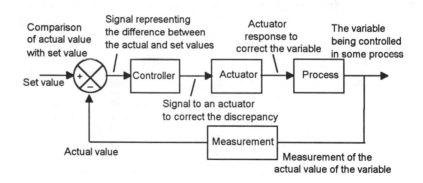

Figure 9.11 *Continuous control*

Figure 9.12 shows the arrangement that might be used with a PLC used to exercise the closed-loop control. It has been assumed that the actuator and the measured values are analogue and thus require conversion to digital; analogue-to-digital and digital-to-analogue units have thus been shown.

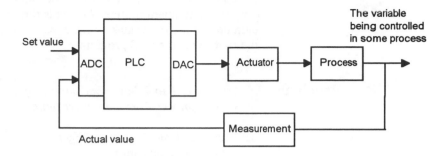

Figure 9.12 *PLC for closed-loop control*

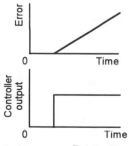

Figure 9.13 *Proportional control*

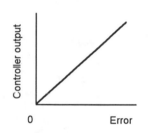

Figure 9.14 *Integral control*

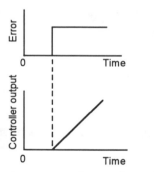

Figure 9.15 *Derivative control*

With *proportional control* the controller gives an output to the actuator which is proportional to the difference between the actual value and the set value of the variable, i.e. the error (Figure 9.13). Such a form of control can be given by a PLC with basic arithmetic facilities. The set value and the actual values are likely to be analogue and so are converted to digital and then the actual value is subtracted from the set value and the difference multiplied by some constant, the proportional constant K_P, to give the output, which after conversion to analogue is the correction signal applied to the actuator:

controller output = $K_P \times$ error

Proportional control has a disadvantage in that, because of time lags inherent in the system, the correcting signal applied to the actuator tends to cause the variable to oscillate about the set value. What is needed is a correcting signal which is reduced as the variable gets close to the set value. This is obtained by *PID control*, the controller giving a correction signal which is computed from a proportional element, the P term, an element which is related to previous values of the variable, the integral I term, and an element related to the rate at which the variable is changing, the derivative D term. With integral control the controller output is proportional to the integral of the error with time, i.e. the area under the error-time graph (Figure 9.14).

controller output = $K_I \times$ integral of error with time

With derivative control the controller output is proportional to the rate at which the error is changing, i.e. the slope of the error-time graph (Figure 9.15):

controller output = $K_D \times$ rate of change of error

The term *tuning* is used for determining the optimum values of K_P, K_I and K_D to be used for a particular control system.

Many PLCs provide the PID calculation to determine the controller output as a standard routine. All that is then necessary is to pass the desired parameters, i.e. the values of K_P, K_I and K_D, and input/output locations to the routine via the PLC program.

Problems

Questions 1 to 7 have four answer options A, B, C or D. Choose the correct answer from the answer options.

Problems 1 and 2 refer to Figure 9.16 which shows two formats used for the move operation.

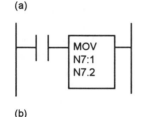

(a)

(b)

Figure 9.16 *Problems 1 and 2*

1 Decide whether each of these statements is True (T) or False (F).

In Figure 9.16(a), the program instruction is to:
(i) Move the value in S to D, leaving S empty.
(ii) Copy the value in S and put it in D.

A (i) T (ii) T
B (i) T (ii) F
C (i) F (ii) T
D (i) F (ii) F

2 Decide whether each of these statements is True (T) or False (F).

In Figure 9.16(b), the program instruction is to:
(i) Move the value in N7.1 to N7.2, leaving N7.1 empty.
(ii) Copy the value in N7.1 and put it in N7.2.

A (i) T (ii) T
B (i) T (ii) F
C (i) F (ii) T
D (i) F (ii) F

Problems 3 and 4 refer to Figure 9.17 which shows two versions of a ladder rung involving a comparison.

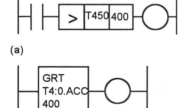

(a)

(b)

Figure 9.17 *Problems 3 and 4*

3 Decide whether each of these statements is True (T) or False (F).

In Figure 9.17(a), the program instruction is to give an output:
(i) When the accumulated time in timer T450 exceeds a value of 400.
(ii) Until the accumulated time in timer T450 reaches a value of 400.

A (i) T (ii) T
B (i) T (ii) F
C (i) F (ii) T
D (i) F (ii) F

4 Decide whether each of these statements is True (T) or False (F).

In Figure 9.17(b), the program instruction is to give an output:
(i) When the accumulated time in timer T4:0 exceeds a value of 400.

(ii) Until the accumulated time in timer T4:0 reaches a value of 400.

A (i) T (ii) T
B (i) T (ii) F
C (i) F (ii) T
D (i) F (ii) F

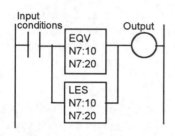

Figure 9.18 *Problem 5*

5 Decide whether each of these statements is True (T) or False (F).

In Figure 9.18 the program instruction is, when the input conditions are met, to give an output when the data:
(i) In N7:10 equals that in N7:20.
(ii) In N7:10 is less than that in N7:20.
A (i) T (ii) T
B (i) T (ii) F
C (i) F (ii) T
D (i) F (ii) F

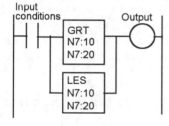

Figure 9.19 *Problem 6*

6 Decide whether each of these statements is True (T) or False (F).

In Figure 9.19, the program instruction is to give, when the input conditions are met, an output when:
(i) The data in N7:10 is not equal to that in N7:20.
(ii) The data in N7.10 is greater or less than that in N7.20.

A (i) T (ii) T
B (i) T (ii) F
C (i) F (ii) T
D (i) F (ii) F

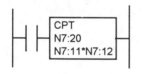

Figure 9.20 *Problem 7*

7 In Figure 9.20, when the input conditions are met, the program instruction is to give in N7.20:

A The sum of the data in N7:11 and N7:12.
B The product of the data in N7:11 and N7:12.
C The difference between the data in N7:11 and N7:12.
D The value given by dividing the data in N7:11 by that in N7:12.

8 Devise ladder programs for systems that will carry out the following tasks:
(a) Switch on a pump when the water level in a tank rises above 1.2 m and switch it off when it falls below 1.0 m.
(b) Switch on a pump, then 100 s later switch on a heater, then a further 30 s later switch on the circulating motor.

10 Designing programs

This chapter considers how programs can be designed and extends the examples given in previous chapters to show programs developed to complete specific tasks. These include tasks which involve temperature control and a number involving pneumatic valves.

10.1 Program development

Whatever the language in which a program is to be written, a systematic approach to the problem can improve the chance of high quality programs being generated in as short a time as possible. A systematic design technique is likely to involve the following steps:

1 A definition of what is required with the inputs and outputs specified.

2 A definition of the algorithm to be used. An algorithm is a step-by-step sequence which defines a method of solving the problem. This can often be shown by a flow chart or written in pseudocode, this involving the use of the words BEGIN, DO, END, IF-THEN-ELSE, WHILE-DO.

3 The algorithm is then translated into instructions that can be inputted to the PLC.

4 The program is then tested and debugged.

5 The program is documented so that any person using or having to modify the program at a later date understands how the program works.

10.1 Flow charts and pseudocode

Consider how the following program operations can be represented by flow charts and pseudocode and then programmed using ladder programming:

1 *Sequential*
Consider a sequence when event A has to be followed by event B. Figure 10.1 shows how this can be represented by a flow chart. In pseudocode this is written as:

```
BEGIN A
      DO A
END A
BEGIN B
      DO B
END B
```

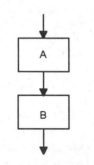

Figure 10.1 *Sequence*

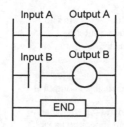

Figure 10.2 *Sequence*

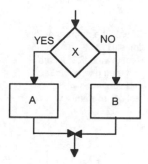

Figure 10.2 *Condition*

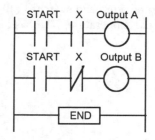

Figure 10.3 *Condition*

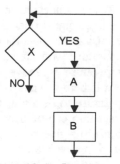

Figure 10.4 *Looping*

A sequence can be translated into a ladder program in the way shown in Figure 10.2. When the start input occurs then output A happens. When A is completed it operates input B and results in output B occurring.

2 *Conditional*

Figure 10.3 shows the flow chart for when A or B is to happen if a particular condition X being YES or NO occurs. The pseudocode to describe this involves the words IF-THEN-ELSE-ENDIF.

IF X
THEN
      BEGIN A
      DO A
      END A
ELSE
      BEGIN B
      DO B
      END B
ENDIF X

Such a condition can be represented by the ladder diagram shown in Figure 10.3. When the start input occurs, the output will be A if there is an input to X, otherwise the output is B.

3 *Looping*

A loop is a repetition of some element of a program, the element being repeated as long as some condition prevails. Figure 10.4 shows how this can be represented by a flow chart. As long as condition X is realised then the sequence A followed by B occurs and is repeated. When X is no longer realised then the program continues and the looping through A and B ceases. In pseudocode this can be represented by using the words WHILE-DO-ENDWHILE:

WHILE X
      BEGIN A
      DO A
      END A
      BEGIN B
      DO B
      END B
ENDWHILE X

Figure 10.5(a) shows how this can be represented by a ladder diagram and using an internal relay.

Where a loop has to be repeated for a particular number of times, a counter can be used, receiving an input pulse each time a loop occurs and switching out of the loop sequence when the required number of loops has been completed (Figure 10.5(b)).

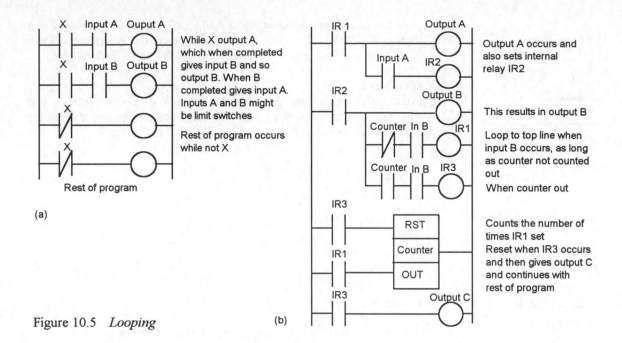

Figure 10.5   *Looping*

10.2 Temperature control

Consider the task of using a PLC as an on–off controller for a heater in the control of temperature in some enclosure. The heater is to be switched on when the temperature falls below the required temperature and switched off when the temperature is at or above the required temperature. The basic algorithm might be considered to be:

        IF temperature below set value
        THEN
                DO switch on heater
        ELSE
                DO switch off heater
        ENDIF

The sensor used for the temperature might be a thermocouple, a thermistor or integrated chip (see section 2.1.5). When connected in an appropriate circuit, the sensor will give a suitable voltage signal related to the temperature. This voltage can be compared, using an operational amplifier, with the voltage set for the required temperature with the result that a high output signal is given when the temperature is above the required temperature and a low output signal when it is below. Thus when the temperature falls from above the required temperature to below it, the signal switches from a high to a low value. This transition can be used as the input to a PLC. The PLC can then be programmed to give an output when there is a low input and this output used to switch on the heater. Figure 10.6 shows the arrangement that might be used and a Mitsubishi ladder program. The input from the operational amplifier has been connected to the input port with the address X400. This input has contacts which are normally closed. When the input goes high, the contacts open. The output is taken from the output port with the address Y430. Thus there is an output when the input is low and no output when the input is high.

Figure 10.7 shows the key sequence that would be used with a graphic programmer to enter the program given in Figure 10.6.

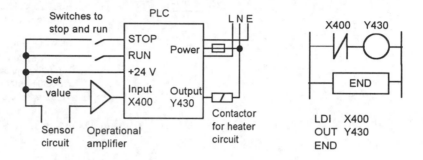

Figure 10.6 *Temperature control*

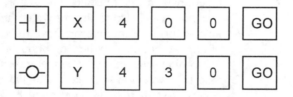

Figure 10.7 *Entering program graphically*

In Siemens format the program given in Figure 10.6 could be as shown in Figure 10.8 and in Allen Bradley form as shown in Figure 10.9. To illustrate how such a program might be entered using a computer, Figure 10.5 also shows the function keys that would be used to enter the program with the Allen Bradley software loaded. At each instant in the program a screen displays prompts and lists the function keys and their significance.

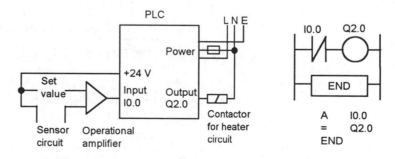

Figure 10.8 *Temperature control*

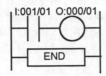

Offline programming selected F1
Mode selected: Create file F6
File name entered: TEMP, enter key pressed
Select the processor: F2 pressed until
relevant one on screen.
Create file F1
Monitor file F8. This shows the file as just
consisting of the END rung.
Edit F10. This is to enable the file to be modified.
Append rung and append instruction F3 pressed
twice.
-||- F1 key pressed.
i;001/01 entered, enter key pressed.
-( )- F3 key pressed.
o;000/01 entered, enter key pressed.
To accept the program Esc is pressed.

Figure 10.9 *Allen Bradley program*

Consider a more complex temperature control task involving a domestic central heating system (Figure 10.10). The central heating boiler is to be thermostatically controlled and supply hot water to the radiator system in the house and also to the hot water tank to provide hot water from the taps in the house. Pump motors have to be switched on to direct the hot water from the boiler to either, or both, of the radiator and hot water systems according to whether the temperature sensors for the room temperature and the hot water tank indicate that the radiators or tank need heating. The entire system is to be controlled by a clock so that it only operates for certain hours of the day. Figure 10.11 shows how a Mitsubishi PLC, and Figure 10.12 a Siemens PLC, might be used.

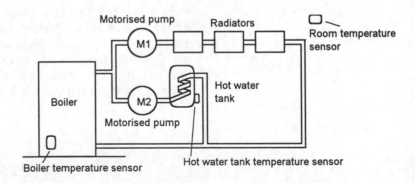

Figure 10.10 *Central heating system*

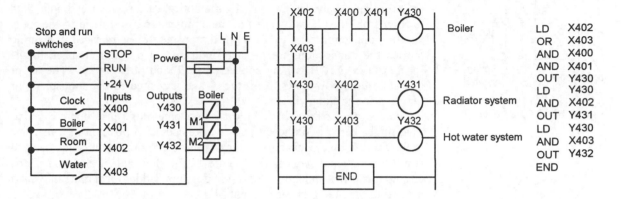

Figure 10.11 *Central heating system*

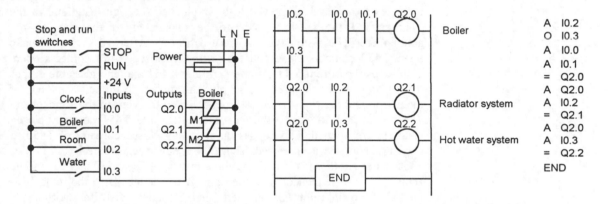

Figure 10.12 *Central heating system*

The boiler, output Y430/Q2.0, is switched on if X400/I0.0 and X401/ I0.1 and either X402/I0.2 or X403/I0.3 are switched on. This means if the clock switched is on, the boiler temperature sensor gives an on input, and either the room temperature sensor or the water temperature sensors give on inputs. The motorised valve M1, output Y431/Q2.1, is switched on if the boiler, Y430/Q2.0, is on and if the room temperature sensor X402/I0.2 gives an on input. The motorised valve M2, output Y432/Q2.2, is switched on if the boiler, Y430/Q2.0, is on and if the water temperature sensor gives an on input.

10.3 Valve sequencing

Consider tasks involving directional control valves (see section 2.2.2 for an introductory discussion). Directional control valves (see Section 2.2.2) are specified in terms of the number of ports and number of control positions they have. Figure 10.13 shows a 4/2 valve; when the pushbutton is pressed, port A is connected to T, the symbol of a vent to the atmosphere or return

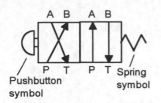

Figure 10.13 *4/2 valve*

of hydraulic fluid to the sump, and port P is connected to P, the symbol for the pressure supply. Hence, pushing the button causes air or hydraulic fluid to flow from P to B and vented to the atmosphere or returned to the sump via the connection of A to T. When the pushbutton is released, the spring pushes the connections to the state indicated in the box attached to the spring. Thus port B is now connected to T and port P to A. The air or hydraulic fluid now flows from P to A and is vented to the atmosphere or returned to the sump via B.

Consider the task of obtaining cyclic movement of a piston in a cylinder. This might be to periodically push workpieces into position in a machine tool with another similar, but out of phase, arrangement being used to remove completed workpieces. Figure 10.14 shows the valve and piston arrangement that might be used, a possible ladder program and chart indicating the timing of each output.

Consider both timers set for 10 s. When the start contacts X400 are closed, timer T450 starts. Also there is an output from Y431. The output Y431 is one of the solenoids used to actuate the valve. When it is energised it causes the pressure supply P to be applied to the right-hand end of the cylinder and the left-hand side to be connected to the vent to the atmosphere. The piston thus moves to the left. After 10 s, the normally open T450 contacts close and the normally closed T450 contacts open. This stops the output Y431, starts the timer T451 and energises the output Y430. As a result, the pressure supply P is applied to the left-hand side of the piston and the right-hand side connected to the vent to the atmosphere. The piston now moves to the left. After 10 s, the T451 normally closed contacts are opened. This causes the normally closed contacts of T450 to close and so Y431 is energised. Thus the sequence repeats itself.

Consider another task involving three pistons A, B and C that have to be actuated in the sequence: A to the right, A to the left, B to the right, B to the left, C to the right, C to the left (such a sequence is often written A+, A–, B+, B–, C+, C–). Figure 10.15(a) illustrates the valves that might be used and Figures 10.15(b) and (c) ladder programs that might be used involving timers. An alternative would involve the use of a shift register.

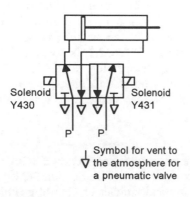

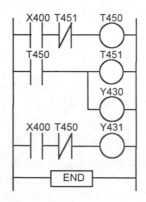

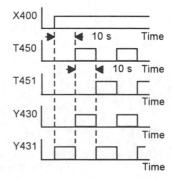

Figure 10.14 *Cyclic movement of a piston*

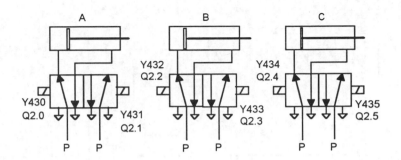

(a)

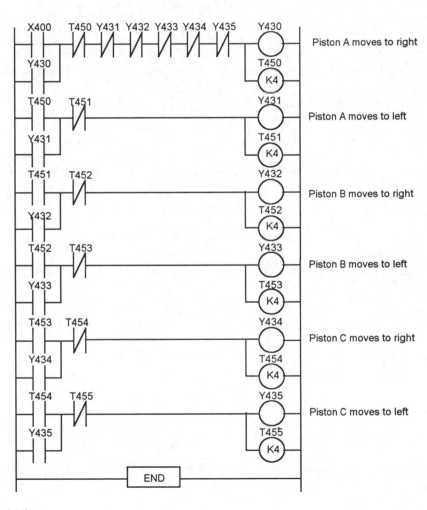

(b) *Mitsubishi format*

Figure 10.15 (*Continued on next page*)

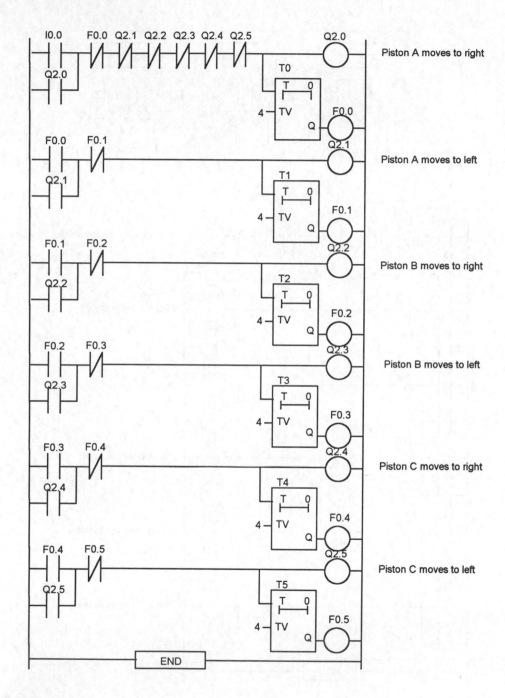

(c) *Siemens format*

Figure 10.15 *Sequencing*

X400/I0.0 is the start switch. When it is closed there is an output from Y430/Q2.0 and the timer T450/T0 starts. The start switch is latched by the output. Piston A moves to the right. After the set time, K = 4, the normally closed timer T450/internal relay F0.0 contacts open and the normally open timer T450/internal relay F0.0 contacts close. This switches off Y430/Q2.0 and energises Y431/Q2.1 and starts timer T451/T1. Piston A moves to the left. In rung 2, the T450/internal relay F0.0 contacts are latched and so the output remains on until the set time has been reached. When this occurs the normally closed timer T451/internal relay F0.1 contacts open and the normally open T451/internal relay F0.1 contacts close. This switches off Y431/Q2.1 and energises Y432/Q2.2 and starts timer T452/T2. Piston B moves to the right. Each succeeding rung activates the next solenoid. Thus in sequence, each of the outputs is energised.

The program instruction list, in the Mitsubishi format, for the above program is:

LD	X400	Start switch
OR	Y430	
ANI	T450	
ANI	Y431	
ANI	Y432	
ANI	Y433	
ANI	Y434	
ANI	Y435	
OUT	Y430	Piston A moves to right
OUT	T450	Timer 450 starts
LD	T450	
OR	Y431	
ANI	T451	
OUT	Y431	Piston A moves to left
OUT	T451	Timer T451 starts
LD	T451	
OR	Y432	
ANI	T452	
OUT	Y432	Piston B moves to right
OUT	T452	Timer T452 starts
LD	T452	
OR	Y433	
ANI	T453	
OUT	Y433	Piston B moves to left
OUT	T453	Timer T453 starts
LD	T453	
OR	Y434	
ANI	T454	
OUT	Y434	Piston C moves to right
OUT	T454	Timer T454 starts
LD	T454	
OR	Y435	
ANI	T455	

OUT	Y435	Piston C moves to left
OUT	T455	Timer T455 starts
END		

10.4 Car park barriers

As another example, consider the use of pneumatic valves to operate car park barriers. The in-barrier is to be opened when the correct money is inserted in the collection box, the out-barrier is to open when a car is detected at that barrier. Figure 10.16 shows the type of system that might be used. The valves used to operate the barriers have a solenoid to obtain one position and a return spring to give the second position. Thus when the solenoid is not energised, the position given is that obtained by the spring. The valves are used to cause the pistons to move. When the pistons move upwards the movement causes the barrier to rotate about its pivot and so lift. When a piston retracts, under the action of the return spring, the barrier is lowered. When a barrier is down it trips a switch and when up it trips a switch, these switches being used to give inputs indicating when the barrier is down and up. Sensors are used to indicate when the correct money has been inserted in the collection box for a vehicle to enter and to sense when a vehicle has approached the exit barrier.

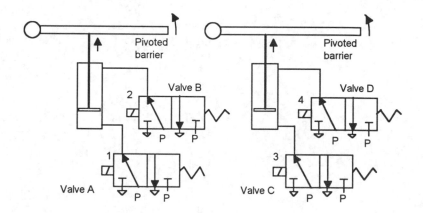

Figure 10.16 *Valve–piston system*

Figure 10.17 shows the form a ladder program could take, (a) being the Mitsubishi program and (b) the Siemens program. The output Y430/Q2.0 to solenoid 1 to raise the entrance barrier is given when the output from the coin box sensor gives the X400/I0.0 input. The Y430/Q2.0 is latched and remains on until the internal relay M100/F0.1 opens. The output will also not occur if the barrier is in the process of being lowered and there is the output Y431/Q2.1 to solenoid 2. The timer T450/T1 is used to hold the barrier up for 10 s, being started by input X402/I0.2 from a sensor indicating the barrier is up. At the end of that time, the output Y431/Q2.1 is switched on, activates solenoid 2 and lowers the barrier. The exit barrier is raised by the output Y432/Q2.2 to solenoid 3 when a sensor detects a car and gives the input X401/I0.1. When the barrier is up a timer T451/T2 is used to hold the barrier up for 10 s, being started by input X404/I0.4 from a sensor indicating the barrier is up. At the end of the time, the output Y433/Q2.3 is switched on, activating solenoid 4 and lowering the barrier.

The program instruction lists for the Mitsubishi and Siemens formats of the program are:

Mitsubishi		*Siemens*	
LD	X400	A	I0.0
OR	Y430	O	Q2.0
ANI	M100	AN	F0.1
ANI	Y431	AN	Q2.1
OUT	Y430	=	Q2.0
LD	X401	A	I0.1
OUT	T450	LKT	10.2
K	10	SR	T0
LD	T450	A	T0
OUT	M100	=	Q2.0
LD	M100	A	F0.1
OR	Y431	O	Q2.1
ANI	X402	AN	I0.2
ANI	Y430	AN	Q2.0
OUT	Y431	=	Q2.1
LD	X403	A	I0.3
OR	Y432	O	Q2.2
ANI	M101	AN	F0.2
ANI	Y433	AN	Q2.3
OUT	Y432	=	Q2.2
LD	X404	A	I0.4
OUT	T451	LKT	10.2
K	10	SR	T1
LD	T451	A	T1
OUT	M101	=	F0.2
LD	M101	A	F0.2
OR	Y433	O	Q2.3
ANI	X405	AN	I0.5
ANI	Y432	AN	Q2.2
OUT	Y433	=	Q2.3
END		END	

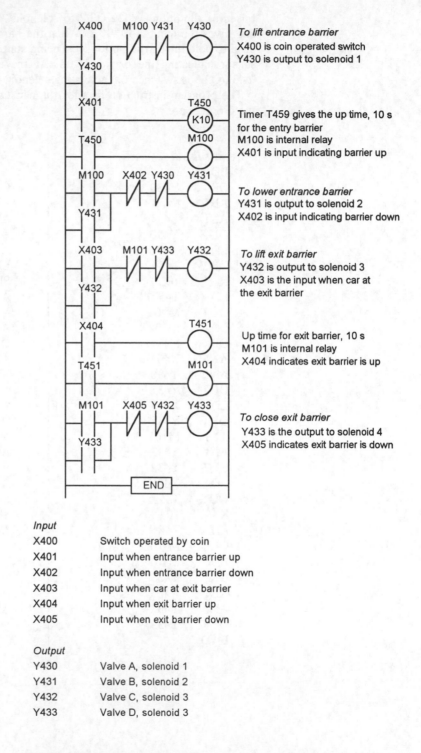

Input
X400	Switch operated by coin
X401	Input when entrance barrier up
X402	Input when entrance barrier down
X403	Input when car at exit barrier
X404	Input when exit barrier up
X405	Input when exit barrier down

Output
Y430	Valve A, solenoid 1
Y431	Valve B, solenoid 2
Y432	Valve C, solenoid 3
Y433	Valve D, solenoid 3

Figure 10.17 *(a) Car barrier program, Mitsubishi format*

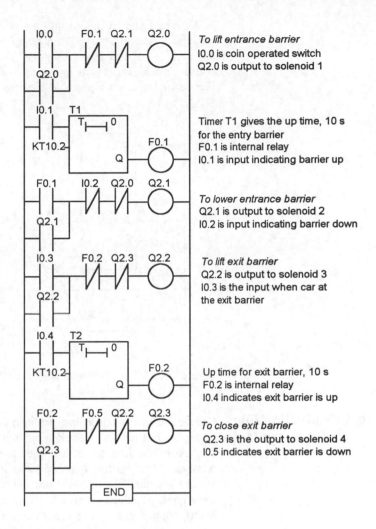

I0.0 is coin operated switch
Q2.0 is output to solenoid 1

Timer T1 gives the up time, 10 s
for the entry barrier
F0.1 is internal relay
I0.1 is input indicating barrier up

To lower entrance barrier
Q2.1 is output to solenoid 2
I0.2 is input indicating barrier down

To lift exit barrier
Q2.2 is output to solenoid 3
I0.3 is the input when car at
the exit barrier

Up time for exit barrier, 10 s
F0.2 is internal relay
I0.4 indicates exit barrier is up

To close exit barrier
Q2.3 is the output to solenoid 4
I0.5 indicates exit barrier is down

Input

I0.0	Switch operated by coin
I0.1	Input when entrance barrier up
I0.2	Input when entrance barrier down
I0.3	Input when car at exit barrier
I0.4	Input when exit barrier up
I0.5	Input when exit barrier down

Output

Q2.0	Valve A, solenoid 1
Q2.1	Valve B, solenoid 2
Q2.2	Valve C, solenoid 3
Q2.3	Valve D, solenoid 3

Figure 10.17 *(b) Car barrier program, Siemens format*

We could add to this program a system to keep check of the number of vehicles in the car park, illuminating a sign to indicate 'Spaces' when the car park is not full and a sign 'Full' when there are no more spaces. This could be achieved by using an up and down counter (see section 7.2.1). Figure 10.18 shows a possible Siemens ladder program.

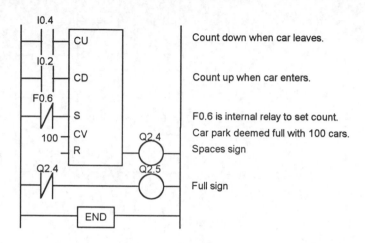

Figure 10.18 *Car park with spaces or full*

10.5 Production line control

Consider a production line problem involving a conveyor being used to transport bottles to a packaging unit, the items being loaded onto the conveyor, checked to ensure they are full, capped and then the correct number (4) of bottles being packed in a container. The required control actions are thus: if a bottle is not full the conveyor is stopped; activation of the capping machine when a bottle is at the required position, the conveyor being stopped during this time; count four bottles and activate the packing machine, the conveyor being stopped if another bottle comes to the packing point at that time; sound an alarm when the conveyor is stopped. The detection of whether a bottle is full could be done with a photoelectric sensor which could then be used to activate a switch (X402/I0.2 input). The presence of a bottle for the capping machine could also be by means of a photoelectric sensor (X403/I0.3 input). The input to the counter to detect the four bottles could be also from a photoelectric sensor (X404/I0.4 input). The other inputs could be start (X400/I0.0 input) and stop (X401/I0.1 input) switches for the conveyor and a signal (X405/I0.5 input) from the packaging machine as to when it is operating, having got four bottles and so is not ready for any further caps. Figure 10.19 shows a possible ladder program in Mitsubishi format, and Figure 10.20 in Siemens format, that could be used.

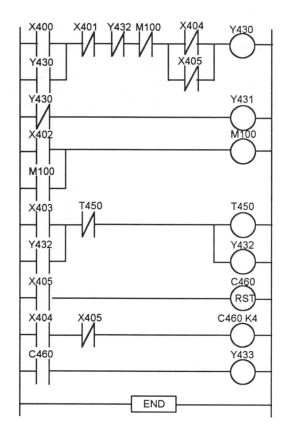

Y430 is the output to the conveyor. X400 is the start button, X401 the stop button. The conveyor is stopped by Y232, M100, X404 or X405 being activated.

Y431 is the output to the alarm. It is triggered when the conveyor stops.

M100 is an internal relay activated by X402 closing when a bottle is not full. It then stops the conveyor.

T450 is a timer which stops the conveyor for time taken to cap the bottle. Y432 energises the capping machine and stops the conveyor.

Reset for the counter when packaging machine has 4 bottles.

X404 input when bottle detected. X405 opens when packing occurring. 4 bottles counted.

Y433 energises packing machine when C460 has counted 4 bottles.

LD	X400	First rung		LD	X403	Fourth rung
OR	Y430			OR	Y432	
ANI	X401			ANI	T450	
ANI	Y432			OUT	T450	
ANI	M100			K	2	2 s allowed for capping.
LDI	X404			OUT	Y432	
ORI	X405			LD	X405	Fifth rung
ANB				RST	C460	
OUT	Y430			LD	X404	Sixth rung
LDI	Y430	Second rung		ANI	X405	
OUT	Y431			OUT	C460	
LD	X402	Third rung		K	4	Four bottles counted.
OR	M100			LD	C460	Seventh rung
OUT	M100			OUT	Y433	
				END		End rung

Figure 10.19 *Production control program*

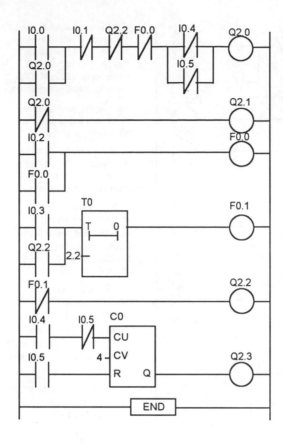

Q2.0 is the output to the conveyor. I0.0 is the start button, I0.1 the stop button. The conveyor is stopped by Q2.2, F0.0, I0.4 or I0.5 being activated.

Q2.1 is the output to the alarm. It is triggered when the conveyor stops.

F0.0 is an internal relay activated by I0.2 closing when a bottle is not full. It then stops the conveyor.

T1 is a timer which stops the conveyor for time taken to cap the bottle.

Q2.2 energises the capping machine and stops the conveyor.

I0.4 input when bottle detected. I0.5 opens when packing occurring. 4 bottles counted.

Q2.3 energises packing machine when counter has counted 4 bottles.

A	I0.0	First rung	A	I0.3	Fourth rung
O.	Q2.0		O.	Q2.2	
AN	I0.1		LKT	2.2	2 s allowed for capping.
AN	Q2.2		SR	T0	
AN	F0.0		A	T0	
(AN	I0.4		=	F0.1	
ON	I0.5		AN	F0.1	Fifth rung
)			=	Q2.2	
=	Q2.0		A	I0.4	Sixth rung
AN	Q2.0	Second rung	AN	I0.5	
=	Q2.1		CU	C0	
A	I0.2	Third rung	LKC	4	Four bottles counted.
O.	F0.0		A	I0.5	
=	F0.0		R	C0	
			=	Q2.3	
			END		End rung

Figure 10.20 *Production control program*

Problems

1 This problem is essentially part of the domestic washing machine program. Devise a ladder program to switch on a pump for 100 s. It is then to be switched off and a heater switched on for 50 s. Then the heater is switched off and another pump is used to empty the water.

2 Devise a ladder program that can be used with a solenoid valve controlled double-acting cylinder, i.e. a cylinder with a piston which can be moved either way by means of solenoids for each of its two positions, and which moves the piston to the right, holds it there for 2 s and then returns it to the left.

3 Devise a ladder program that could be used to operate the simplified task shown in Figure 10.21 for the automatic drilling of workpieces. The drill motor and the pump for the air pressure for the pneumatic valves have to be started. The workpiece has to be clamped. The drill has then to be lowered and drilling started to the required depth. Then the drill has to be retracted and the workpiece unclamped.

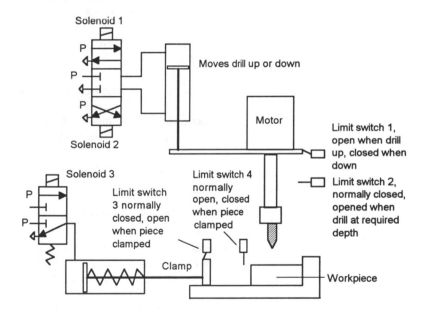

Figure 10.21 *Problem 3*

11 Testing and debugging

The aims of this chapter are to indicate how a PLC system can be tested and faults found. This involves consideration of both the hardware and the software.

11.1 Commissioning and testing

Commissioning of a PLC system involves:

1   Checking that all the cable connections between the PLC and the plant being controlled are complete, safe and to the required specification and meeting local standards.

2   Checking that the incoming power supply matches the voltage setting for which the PLC is set.

3   Checking that all protective devices are set to their appropriate trip settings.

4   Checking that emergency stop buttons work.

5   Checking that all input/output devices are connected to the correct input/output points and giving the correct signals.

6   Loading and testing the software.

11.1.1 Testing inputs and outputs

Input devices, e.g. switches, can be manipulated to give the open and closed contact conditions and the corresponding LED on the input module observed. It should be illuminated when the input is closed and not illuminated when it is open. Failure of an LED to illuminate could be because the input device is not correctly operating, there are incorrect wiring connections to the input module, the input device is not correctly powered or the LED or input module is defective. For output devices that can be safely started, push buttons might have been installed so that each output can be tested.

Another method that can be used to test inputs and outputs is termed *forcing*. This involves software, rather than mechanical switching on or off, being used with instructions from the programming panel to turn off or on inputs/outputs. In order to do this, a PLC has to be switched into the forcing or monitor mode by perhaps pressing a key marked FORCE or selecting that mode on a screen display. For example, Figure 11.1 shows, for a Sprecher+Schuh PLC, the keystrokes that might be used, and the resulting screen display, to force the output Y005 into the on state. Figure 11.2 shows the keys for the forcing of an input X001 into a closed state. Thus if an input is forced and the input LED comes on then we can check that the consequential action of that input being on occurs.

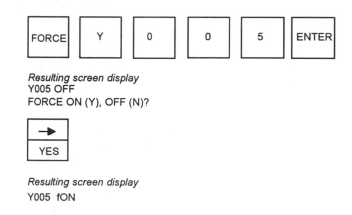

Resulting screen display
Y005 OFF
FORCE ON (Y), OFF (N)?

Resulting screen display
Y005 fON

Figure 11.1 *Forcing an output*

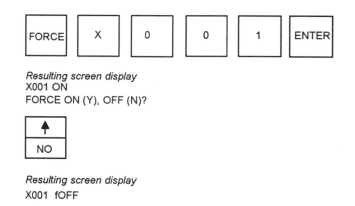

Resulting screen display
X001 ON
FORCE ON (Y), OFF (N)?

Resulting screen display
X001 fOFF

Figure 11.2 *Forcing an input*

11.1.2 Testing software

Most PLCs contain some software checking program. This checks through the installed program for incorrect device addresses, and provides a list on a screen or as a printout of all the input/output points used, counter and timer settings, etc. with any errors detected. For example, there might be a message for a particular output address that it is used as an output more than once in the program, a timer or counter is being used without a preset value, a counter is being used without a reset, etc.

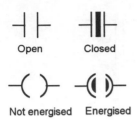

Figure 11.3 *Monitor mode symbols*

11.1.3 Simulation

Many PLCs are fitted with a simulation unit which reads and writes information directly into the input/output memory and so simulates the actions of the inputs and outputs. The installed program can thus be run and inputs and outputs simulated so that they, and all preset values, can be checked. To carry out this type of operation the terminal has to be placed in the correct mode. For Mitsubishi this is termed the monitor mode, for Siemens the test mode, for Telemecanique the debug mode, for Sprecher+Schuh the read mode/monitor mode.

With Mitsubishi in the monitor mode, Figure 11.3 shows how inputs appear when open and closed, and output when not energised and energised. The display shows a selected part of the ladder program and what happens as the program proceeds. Thus at some stage in a program the screen might appear in the form shown in Figure 11.4(a). For rung 12, with inputs to X400, X 401 and X402, but not M100, there is no output from Y430. For rung 13, the timer T450 contacts are closed, the display at the bottom of the screen indicating that there is no time left to run on T450. Because Y430 is not energised the Y430 contacts are open and so there is no output from Y431. If we now force an input to M100 then the screen display changes to that shown in Figure 11.4(b). Now Y430, and consequently Y431, come on.

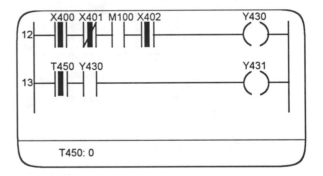

(a)

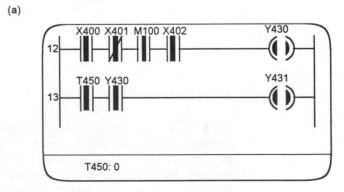

(b)

Figure 11.4 *Ladder program monitoring*

11.2 Fault finding

With any PLC controlled plant, by far the greater percentage of the faults are likely to be with sensors, actuators and wiring rather than within the PLC itself. Of the faults within the PLC, most are likely to be in the input/output channels or power supply than in the CPU.

As an illustration of a fault, consider a single output device failing to turn on though the output LED is on. If testing of the PLC output voltage indicates that it is normal then the fault might be a wiring fault or a device fault. If checking of the voltage at the device indicates the voltage there is normal then the fault is the device. As another illustration, consider all the inputs failing. This might be as a result of a short circuit or earth fault with an input and a possible procedure to isolate the fault is to disconnect the inputs one by one until the faulty input is isolated. An example of another fault is if the entire system stops. This might be a result of a power failure, or someone switching off the power supply, or a circuit breaker tripping.

Many PLCs provide built-in fault analysis procedures which carry out self-testing and display fault codes, with possibly a brief message, which can be translated by looking up the code in a list to give the source of the fault and possible methods of recovery. For example, the fault code may indicate that the source of the fault is in a particular module with the method of recovery given as replace that module or perhaps switch the power off and then on.

11.2.1 Fault detection techniques

The following are some of the common fault detection techniques used:

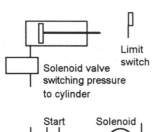

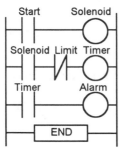

Figure 11.5 *Watchdog timer*

1   *Timing checks*
    The term *watchdog* is used for a timing check that is carried out by the PLC to check that some function has been carried out within the normal time. If the function is not carried out within the normal time then a fault is assumed to have occurred and the watchdog timer trips, setting of an alarm and perhaps closing down the PLC.

    Within a program additional ladder rungs might be included so that when a function starts a timer is started. If the function is completed before the time runs out then the program continues but if not the program uses the jump command to move to a special set of rungs which triggers off an alarm and perhaps stops the system. Figure 11.5 shows an example of a watchdog timer which might be used with the movement of a piston in a cylinder. When the start switch is closed the solenoid of a valve is energised and causes the piston in the cylinder to start moving. It also starts the timer. When the piston is fully extended it opens a limit switch and stops the timer. If the time taken for the piston to move and switch off the timer is greater than the preset value used for the timer, the timer sets off the alarm.

2   *Replication*
    Where there is concern regarding safety in the case of a fault developing, checks may be constantly used to detect faults. One

technique is *replication checks* which involves duplicating, i.e. replicating, the PLC system. This could mean that the system just repeats every operation twice and if it gets the same result it is assumed there is no fault. This procedure can detect transient faults. A more expensive alternative is to have duplicate PLC systems and compare the results given by the two systems. In the absence of a fault the two results should be the same, a fault showing up as a difference.

3   *Expected value checks*
    Software errors can be detected by checking whether an expected value is obtained when a specific input occurs. If the expected value is not obtained then a fault is assumed to be occurring.

11.2.2 Program storage

Applications programs may be loaded into battery-backed-up RAM in a PLC. A failure of the battery supply means a complete loss of the stored programs. An alternative to storing applications programs in battery-backed RAM is to use EPROM. This form of memory is secure against the loss of power. Against the possibility of memory failure occurring in the PLC and loss of the stored application program, a back-up copy of each application program should be kept. If the program has been developed using a computer, the back-up may be on a floppy disk or a hard disk. Otherwise the back-up may be on an EPROM cartridge. The program can then again be downloaded into the PLC without it having to be rewritten.

11.3 System documentation

The documentation is the main guide used by the everyday users and for troubleshooting and fault finding with PLCs. It thus needs to be complete and in a form that is easy to follow. The documentation for a PLC installation should include:

1   A description of the plant.
2   Specification of the control requirements.
3   Details of the programmable logic controller.
4   Electrical installation diagrams.
5   Lists of all input and output connections.
6   Application program with full commentary on what it is achieving.
7   Software back-ups.
8   Operating manual, including details of all start up and shut down procedures and alarms.

Problems

Questions 1 to 6 have four answer options A, B, C or D. Choose the correct answer from the answer options.

1   Decide whether each of these statements is True (T) or False (F).

The term forcing when applied to a PLC input/output means using a program to:
(i) Turn on or off inputs/outputs.
(ii) Check that all inputs/outputs give correct responses when selected.

A (i) T (ii) T
B (i) T (ii) F
C (i) F (ii) T
D (i) F (ii) F

2 Decide whether each of these statements is True (T) or False (F).

The term watchdog when applied to a PLC means a checking mechanism that:
(i) Excessive currents are not occurring.
(ii) Functions are carried out within prescribed time limits.

A (i) T (ii) T
B (i) T (ii) F
C (i) F (ii) T
D (i) F (ii) F

3 Decide whether each of these statements is True (T) or False (F).

When a PLC is in monitor/test/debug mode it:
(i) Enables the operation of a program to be simulated.
(ii) Carries out a fault check.

A (i) T (ii) T
B (i) T (ii) F
C (i) F (ii) T
D (i) F (ii) F

Figure 11.6 *Problem 4*

4 When a PLC is in monitor/test/debug mode and the symbol shown in Figure 11.6 occurs, it means that an input is:

A Defective.
B Correctly operating.
C On.
D Off.

5 Decide whether each of these statements is True (T) or False (F).

Failure of an input sensor or its wiring, rather than failure of an LED or in the PLC input channel, will show as:
(i) The input LED not coming on.
(ii) Forcing of that input making the input LED come on.

A (i) T (ii) T
B (i) T (ii) F
C (i) F (ii) T
D (i) F (ii) F

6 A single output device fails to turn on when the output LED is on. The voltage at the output is tested and found normal but the voltage at the device is found to be absent. The fault is:

A Faulty wiring.
B A faulty output device.
C A fault in the PLC.
D A fault in the program.

7 Explain how, using forcing, the failure of an input sensor or its wiring can be detected.

8 Suggest possible causes of a complete stoppage of the control operation and the PLC with the power on lamp off.

9 Suggest possible causes of an output LED being on but the output device failing to turn on.

Appendix: Number systems

The number system used for everyday calculations is the *denary* or *decimal system*. This is based on the use of the 10 digits: 0, 1, 2, 3, 4, 5, 6, 7, 8, 9. With a number represented by this system, the digit position in the number indicates the weight attached to each digit, the weight increasing by a factor of 10 as we proceed from right to left. Hence we have:

...	10^3	10^2	10^1	10^0
	thousands	hundreds	tens	units

The *binary system* is based on just two digits: 0 and 1. These are termed *binary digits* or *bits*. When a number is represented by this system, the digit position in the number indicates the weight attached to each digit, the weight increasing by a factor of 2 as we proceed from right to left. Hence we have:

...	2^3	2^2	2^1	2^0
	bit 3	bit 2	bit 1	bit 0

The bit 0 is termed the *least significant bit* (LSB) and the highest bit the *most significant bit* (MSB). For example, with the binary number 1010, the least significant bit is the bit at the right-hand end of the number and so is 0. The most significant bit is the bit at the left-hand end of the number and so is 1. When converted to a denary number we have, for the 1010:

	2^3	2^2	2^1	2^0
	bit 3	bit 2	bit 1	bit 0
	MSB			LSB
Binary	1	0	1	0
Denary	$2^3 = 8$	0	$2^1 = 2$	0

Thus the denary equivalent is 10. The conversion of a binary number to a denary number thus involves the addition of the powers of 2 indicated by the number.

The conversion of a denary number to a binary number involves looking for the appropriate powers of 2. We can do this by successive divisions by

2, noting the remainders at each division. Thus if we have the denary number 31:

$31 \div 2 = 15$ remainder 1 This gives the LSB
$15 \div 2 = 7$ remainder 1
$7 \div 2 = 3$ remainder 1
$3 \div 2 = 1$ remainder 1 This gives the MSB

The binary number is thus 1111. The first division gives the least significant bit because we have just divided the 31 by 2, i.e. 2^1 and found 1 left over for the 2^0 digit. The last division gives the most significant bit because the 31 has then been divided by 2 four times, i.e. 2^4 and the remainder is 1.

Binary numbers are used in computers because the two states represented by 0 and 1 are easy to deal with switching circuits where they can represent off and on. A problem with binary numbers is that a comparatively small number requires a large number of digits. For example, the denary number 9 which involves just a single digit requires four when written as the binary number 1001. The denary number 181, involving three digits, in binary form is 10110101 and requires eight digits. Because of this, octal or hexadecimal numbers are sometimes used to make numbers easier to handle.

The *octal system* is based on eight digits: 0, 1, 2, 3, 4, 5, 6, 7. When a number is represented by this system, the digit position in the number indicates the weight attached to each digit, the weighting increasing by a factor of 8 as we proceed from right to left. Thus we have:

$$\ldots \qquad 8^3 \qquad 8^2 \qquad 8^1 \qquad 8^0$$

To convert denary numbers to octal we successively divide by 8 and note the remainders. Thus the denary number 15 divided by 8 gives 1 with remainder 7 and thus the denary number 15 is 17 in the octal system. To convert from octal to denary we multiply the digits by the power of 8 appropriate to its position in the number. For example, the octal number 365 is $3 \times 8^2 + 6 \times 8^1 + 5 \times 8^0 = 245$. To convert from binary into octal, the binary number is written in groups of three bits starting with the least significant bit. For example, the binary number 11010110 would be written as:

11 010 110

Each group is then replaced by the corresponding digit 0 to 7. The 110 binary number is 7, the 010 is 2 and the 11 is 3. Thus the octal number is 326. As another example, the binary number 100111010 is:

100 111 010 Binary
 4   7   2   Octal

Octal to binary conversion involves converting each octal digit into its 3-bit equivalent. Thus, for the octal number 21 we have 1 as 001 and 2 as 010:

   2   1   Octal number
010 001 Binary number

and so the binary number is 010001.

The *hexadecimal system (hex)* is based on 16 digits/symbols: 0, 1, 2, 3, 4, 5, 6, 7, 8, 9, A, B, C, D, E, F. When a number is represented by this system, the digit position in the number indicates that the weight attached to each digit increases by a factor of 16 as we proceed from right to left. Thus we have:

$$\ldots \qquad 16^3 \qquad 16^2 \qquad 16^1 \qquad 16^0$$

For example, the decimal number 15 is F in the hexadecimal system. To convert from denary numbers into hex we successively divide by 16 and note the remainders. Thus the denary number 156 when divided by 16 gives 9 with remainder 12 and so in hex is 9C. To convert from hex to denary we multiply the digits by the power of 16 appropriate to its position in the number. Thus hex 12 is $1 \times 16^1 + 2 \times 16^0 = 18$. To convert binary numbers into hexadecimal numbers, we group the binary numbers into fours starting from the least significant number. Thus, for the binary number 1110100110 we have:

11 1010 0110 Binary number
3   R   6   Hex number

For conversion from hex to binary, each hex number is converted to its 4-bit equivalent. Thus, for the hex number 1D we have 0001 for the 1 and 1101 for the D:

   1   D   Hex number
0001 1100 Binary number

Thus the binary number is 0001 1101.

Because the external world tends to deal mainly with numbers in the denary system and computers with numbers in the binary system, there is always the problem of conversion. There is, however, no simple link between the position of digits in a denary number and the position of digits in a binary number. An alternative method that is often used is the *binary coded decimal system (BCD)*. With this system, each denary digit is coded separately in binary. For example, the denary number 15 has the 5 converted into the binary number 0101 and the 1 into 0001:

   1   5   Denary number
0001 0101 Binary number

to give in BCD the number 0001 0101.

Table 1 gives examples of numbers in the denary, binary, octal, hex and BCD systems.

Table 1 *Examples of numbers in different systems*

Denary	Binary	Octal	HEX	BCD
0	00000	0	0	0000 0000
1	00001	1	1	0000 0001
2	00010	2	2	0000 0010
3	00011	3	3	0000 0011
4	00100	4	4	0000 0100
5	00101	5	5	0000 0101
6	00110	6	6	0000 0110
7	00111	7	7	0000 0111
8	01000	10	8	0000 1000
9	01001	11	9	0000 1001
10	01010	12	A	0001 0000
11	01011	13	B	0001 0001
12	01100	14	C	0001 0010
13	01101	15	D	0001 0011
14	01110	16	E	0001 0100
15	01111	17	F	0001 0101
16	10000	20	11	0001 0110
17	10001	21	12	0001 0111

Binary arithmetic

Addition of binary numbers uses the following rules:

$0 + 0 = 0$

$0 + 1 = 1 + 0 = 1$

$1 + 1 = 10$

$1 + 1 + 1 = 11$

Consider the addition of the binary numbers 01110 and 10111.

$$\begin{array}{r} 01110 \\ 10111 \\ \hline \end{array}$$

Sum   100001

For bit 0 in the sum, $0 + 1 = 1$. For bit 1 in the sum, $1 + 1 = 10$ and so we have 0 with 1 carried to the next column. For bit 3 in the sum, $1 + 0 +$ the carried $1 = 10$. For bit 4 in the sum, $1 + 0 +$ the carried $1 = 10$. We continue this through the various bits and end up with the 100001.

Subtraction of binary numbers follows the following rules:

$$0 - 0 = 0$$

$$1 - 0 = 1$$

$$1 - 1 = 0$$

When evaluating $0 - 1$, a 1 is borrowed from the next column on the left containing a 1. The following example illustrates this with the subtraction of 01110 from 11011:

$$\begin{array}{r} 11011 \\ 01110 \\ \hline \text{Difference} \quad 01101 \end{array}$$

For bit 0 we have $1 - 0 = 1$. For bit 1 we have $1 - 1 = 0$. For bit 2 we have $0 - 1$. We thus borrow 1 from the next column and so have $10 - 1 = 1$. For bit 3 we have $0 - 1$, remember we borrowed the 1. Again borrowing 1 from the next column, we then have $10 - 1 = 1$. For bit 4 we have $0 - 0 = 0$, remember we borrowed the 1.

Signed numbers

The binary numbers considered so far contain no indication whether they are negative or positive and are said to be *unsigned*. Since there is generally a need to handle both positive and negative numbers there needs to be some way of distinguishing between them. This can be done by adding a sign bit. When a number is said to be *signed* then the most significant bit is used to indicate the sign of the number, a 0 being used if the number is positive and a 1 if it is negative. Thus for an 8-bit number we have:

XXXX XXXX

Sign bit

When we have a positive number then we write it in the normal way with a 0 preceding it. Thus a positive binary number of 10110 would be written as 010110. A negative number of 10110 would be written as 110110. However, this is not the most useful way of writing negative numbers for ease of manipulation by computers.

A more useful way of writing signed negative numbers is to use the two's complement method. A binary number has two complements, known as the *one's complement* and the *two's complement*. The one's complement of a binary number is obtained by changing all the 1s in the unsigned number into 0s and the 0s into 1s. Thus if we have the binary number 101101 then the one's complement of it is 010010. The two's complement is obtained from the one's complement by adding 1 to the least significant bit of the one's complement. Thus the one's complement of 010010 becomes 010011.

When we have a negative number then, to obtain the signed two's complement, we obtain the two's complement and then sign it with a 1. Consider the representation of the decimal number −6 as a signed two's complement number when the total number of bits is to be eight. We first write the binary number for +6, i.e. 0000110, then obtain the one's complement of 1111001, add 1 to give 1111010, and finally sign it with a 1 to indicate it is negative. The result is thus 11111010.

Unsigned binary number when sign ignored	000 0110
One's complement	111 1001
Add 1	1
Unsigned two's complement	111 1010
Signed two's complement	1111 1010

Table 2 lists some signed two's complements, given to 4 bits, for denary numbers.

Table 2 *Signed two's complements*

Denary number	Signed 2s complement
−5	1011
−4	1100
−3	1101
−2	1110
−1	1111

When we have a positive number then we sign the normal binary number with a 0, i.e. we only write negative numbers in the two's complement form. A consequence of adopting this method of writing negative and positive numbers is that when we add the signed binary equivalent of +4 and −4, i.e. 0000 0100 and 111 1100 we obtain (1)0000 0000 and so zero within the constraints of the number of bits used, the (1) being neglected.

Subtraction of a positive number from a positive number can be considered to be the addition of a negative number to a positive number. Thus we obtain the signed two's complement of the negative number and then add it to the signed positive number. Hence, for the subtraction of the denary number 6 from the denary number 4 we can consider the problem as being (+4) + (−6). Hence we add the signed positive number to the signed two's complement for the negative number.

Binary form of +4	0000 0100
(−6) as signed two's complement	1111 1010
Sum	1111 1110

The most significant bit, i.e. the sign, of the outcome is 1 and so the result is negative. This is the 8-bit signed two's complement for −2.

If we wanted to add two negative numbers then we would obtain the signed two's complement for each number and then add them. Whenever a number is negative we use the signed two's complement, when positive just the signed number.

Problems

1   Convert the following binary numbers to denary numbers:

   (a) 000011, (b) 111111, (c) 001101

2   Convert the following denary numbers to binary numbers:

   (a) 100, (b) 146, (c) 255

3   Convert the following hexadecimal numbers to denary numbers:

   (a) 9F, (b) D53, (c) 67C

4   Convert the following denary numbers to hexadecimal numbers:

   (a) 14, (b) 81, (c) 2562

5   Convert the following hexadecimal numbers to binary numbers:

   (a) E, (b) 1D, (c) A65

6   Convert the following octal numbers to denary numbers:

   (a) 372, (b) 14, (c) 2540

7   Convert the following denary numbers to octal numbers:

   (a) 20, (b) 265, (c) 400

8   Convert the following octal numbers to binary numbers:

   (a) 270, (b) 102, (c) 673

9   Convert the following decimal numbers to BCD equivalents:

   (a) 20, (b) 35, (c) 92

10 Convert the following denary numbers to signed two's complement binary 8-bit format:

(a) –1, (b) –35, (c) –125

11 Convert the following signed two's complement binary 8-bit numbers to their denary equivalents:

(a) 1111 0000, (b) 1100 1001, (c) 1101 1000

Answers

Chapter 1 1 D 2 A 3 C 4 A 5 A 6 C
7 See Figure 1.4.
8 See Figure 1.8 and associated text.
9 See section 1.3.4.
10 2×1024.

Chapter 2 1 A 2 A 3 B 4 D 5 C 6 A
7 A 8 A 9 B 10 C
11 See (a) Figure 2.4, (b) section 2.1.4, (c) section 2.1.3, (d) Figure 2.2.
12 See Figures 2.38 and 2.39.
13 See section 2.2.3.

Chapter 3 1 B 2 A 3 C 4 C 5 A 6 C
7 A 8 D 9 D
10 (a) 0, (b) 1.
11 To detect message corruption.
12 See section 3.5.

Chapter 4 1 A 2 D 3 B 4 C 5 B 6 B
7 B 8 D 9 C 10 A 11 C 12 A
13 B 14 D 15 A 16 C 17 A 18 B
19 D 20 A 21 A 22 B 23 D 24 C
25 D 26 A 27 C 28 C 29 D 30 C
31 A
32 See (a) Figure 4.10, (b) Figure 4.13, (c) Figure 4.27, (d) Figure 4.13, (e) Figure 4.16, (f) Figure 4.6.

Chapter 5 1 D 2 B 3 C 4 A 5 C 6 C
7 A 8 A 9 C 10 D 11 B 12 B
13 B 14 C 15 A 16 A 17 A 18 B
19 A 20 A 21 B
22 See (a) Figure 5.8, (b) Figures 5.9 or 5.10 or 5.16, (c) Figure 5.20.

Chapter 6
1 C	2 B	3 D	4 A	5 D	6 D
7 C	8 C	9 B	10 C	11 A	12 A
13 A	14 B	15 D	16 B	17 A	18 D
19 C					

20 See (a) Figure 6.3, (b) Figure 6.11, (c) Figure 6.14.

Chapter 7
1 C	2 A	3 C	4 B	5 B	6 B
7 B	8 D	9 C	10 A	11 A	12 B
13 B	14 B	15 C	16 D		

17 See (a) Figures 7.3, 7.4, 7.5, (b) Figures 7.9 and 7.10.

Chapter 8
1 D	2 C	3 C	4 D	5 A	6 A
7 C	8 C	9 D			

10 (a) As Figure 8.1/8.3 with a constant input to In 1/X400, so entering a 1 at each shift, (b) As in Figure 8.5 but instead of faulty item, hook with an item, and instead of good item, hooks with no items.

Chapter 9
1 C	2 C	3 B	4 B	5 A	6 A
7 B					

8 Similar to (a) Figure 9.7, (b) Figure 9.8.

Chapter 10
1 See Figure A.1.
2 See Figure A.2.
3 See Figure A.3 for a basic answer.

Chapter 11
1 B	2 C	3 B	4 D	5 A	6 A

7 Power failure, supply off, power tripped.
8 Wiring fault, device fault.

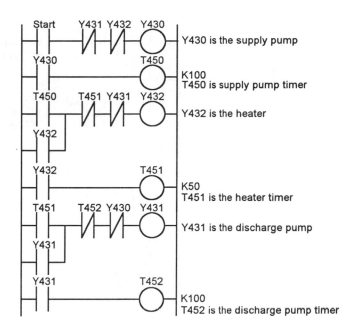

Figure A.1 *Chapter 10, problem 1*

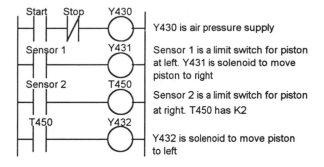

Figure A.2 *Chapter 10, problem 2*

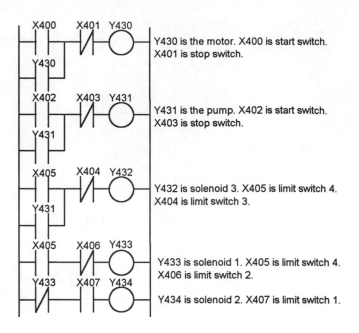

Y430 is the motor. X400 is start switch.
X401 is stop switch.

Y431 is the pump. X402 is start switch.
X403 is stop switch.

Y432 is solenoid 3. X405 is limit switch 4.
X404 is limit switch 3.

Y433 is solenoid 1. X405 is limit switch 4.
X406 is limit switch 2.

Y434 is solenoid 2. X407 is limit switch 1.

Figure A.3 *Chapter 10, problem 3*

Appendix

1 (a) 28, (b) 63, (c) 13
2 (a) 110 0100, (b) 1001 0010, (c) 1111 1111
3 (a) 159, (b) 3411, (c) 1660
4 (a) E, (b) 51, (c) A02
5 (a) 1110, (b) 11101, (c) 1010 0110 0101
6 (a) 250, (b) 12, (c) 1376
7 (a) 24, (b) 411, (c) 620
8 (a) 010 111 000, (b) 001 000 010, (c) 110 111 011
9 (a) 0010 0000, (b) 0011 0101, (c) 1001 0010
10 (a) 1111 1111, (b) 1101 1101, (c) 1000 0011
11 (a) −112, (b) −57, (c) −38

Index